Completion Detection in Asynchronous Circuits

Pallavi Srivastava

Completion Detection in Asynchronous Circuits

Toward Solution of Clock-Related Design Challenges

 Springer

Pallavi Srivastava
Taylor's University
Subang Jaya, Malaysia

ISBN 978-3-031-18396-6   ISBN 978-3-031-18397-3   (eBook)
https://doi.org/10.1007/978-3-031-18397-3

This Springer imprint is published by the registered company Springer Nature Switzerland AG
The registered company address is: Gewerbestrasse 11, 6330 Cham, Switzerland

For Phoebo, my handsome cat.

Preface

Asynchronous design style has experienced a renaissance in recent decades, as the demand for smaller, faster circuits exacerbates the challenges associated with clock distribution, clock skew and other clock-related issues in synchronous circuits. Asynchronous circuits are inherently free from challenges associated with clock signals, as they do not use clock signal(s) to indicate the process completion and instead use handshaking signals to communicate between two asynchronous logic blocks. Using handshaking signals provides an opportunity for the designers of asynchronous circuits to determine the process completion based on the actual input data rather than having to wait for a critical path delay for each computation, as would be the case with a global clock. Hence, asynchronous circuits can utilise the average-case computation delay as they have event-driven characteristics and can indicate data validity as soon as the computation is done, but detecting the completion of the computational process has always been a challenge in such circuits.

Existing completion detection schemes utilise either bundled data or dual-rail data encoding techniques to detect the completion of an event. Most of the researchers prefer dual-rail coding to implement asynchronous circuits even though it needs a large silicon area because it can indicate data validity deterministically. On the other hand, the silicon area required by bundled data protocol is comparable to synchronous design, but it opts for the worst-case delay model to determine the completion of an event. This trade-off between area and speed is eliminated by the completion detection scheme introduced in this book since it can deterministically determine the completion of an event for bundled data circuits. The generic architecture of the deterministic completion detection scheme was developed in accordance with the characteristics of the input data and can be implemented for any asynchronous bundled data circuit that fits a certain set of requirements. When designing a digital circuit, the availability of a generic architecture enables researchers to make an informed choice between synchronous and asynchronous design styles.

Chapter 1 highlighted the difference between synchronous and asynchronous design styles and discusses which design style is better suited for the application for which the electronic circuit is being developed. The background of asynchronous circuit design is also provided in Chap. 1, followed by its fundamentals in Chap. 2. The literature review in Chap. 3 demonstrated various existing completion detection schemes and their shortcomings. While the existing completion detection schemes for bundled data circuits demonstrated the potential for performance improvement, it can be concluded from the literature review that there are still various aspects of bundled data protocol that need to be investigated. Chapter 4 explains the basic architecture of barrel shifter and binary adders, as the proposed generic architecture was validated on these two key modules of an Arithmetic Logic Unit (ALU). The generic architecture of the deterministic completion detection scheme was developed in accordance with the characteristics of the input data and can be implemented for any asynchronous bundled data circuit that fits a certain set of requirements, as discussed in Chap. 5. The generic architecture was validated on barrel shifter and binary adder, by utilising model delays which replicate the actual computation corresponding to the ongoing computation process. Xilinx Vivado® Design Suite was used to conduct the experiments in a simulation environment, and Verilog HDL was used to code the designs. Chapter 6 discusses the optimisation of the barrel shifter and binary adder architecture using deterministic completion detection scheme, followed by their simulation results in Chap. 7. The key contributions of this book are as follows:

- A generic architecture of deterministic completion detection scheme for bundled data circuits
- Pallavi's generic function $\mathcal{G}_p$ to deterministically detect the completion of ongoing computation process
- Reducing the computation delay of bundled data circuits from the worst-case delay to average-case delay with the help of the completion detection scheme, such that they can utilise the event-driven property of asynchronous circuits
- Implementing an adder and a barrel shifter using the proposed design to validate the generic architecture of the deterministic completion detection scheme with the help of the generic function $\mathcal{G}_p$

Subang Jaya, Malaysia   Pallavi Srivastava
August 2022

Acknowledgements

I shall begin with acknowledging Devi Saraswati, the Almighty: without Her guidance, I would never have discovered the right path. I would like to thank my family and friends for their constant support and encouragement. My deepest appreciation goes out to my parents, my husband Abhaya, and sister Trisha. I am indebted to my PhD supervisor, Dr Edwin Chung for his insightful comments and suggestions. Special thanks to Lichi and Phoebo, for keeping me stress-free and allowing me to focus on my work. I would never have been able to reach this stage without their unconditional support and endurance.

This monograph is essentially based on the articles written and published by the author, hence some previously published material is occasionally reused. Even though the work has been modified and rephrased several times for this monograph, the copyright permission from multiple publishers is acknowledged below.

Acknowledging IEEE Publication to reproduce material from the following paper:

- Pallavi Srivastava and Edwin Chung (2021). "An Asynchronous Bundled-Data Barrel Shifter Design That Incorporates a Deterministic Completion Detection Technique", IEEE Transactions on Circuits and Systems II: Express Briefs, https://doi.org/10.1109/TCSII.2021.3126853.

Acknowledging MDPI Publication to reproduce material from the following paper:

- Pallavi Srivastava, Edwin Chung, and Stepan Ozana (2020). "Asynchronous Floating-Point Adders and Communication Protocols: A Survey", Electronics, v (9), no. (10): 1687, https://doi.org/10.3390/electronics9101687.

Contents

Acronyms

ABBA   Asynchronous Bundled-Data Binary Adder
ABBS   Asynchronous Bundled-Data Barrel Shifter
CBS   Conventional Barrel Shifter
CSA   Carry Select Adder
DCDC   Deterministic Completion Detection Circuit
DGU   Delay Generating Unit
DI   Delay Insensitive
EDA   Electronic Design Automation
HDL   Hardware Description Language
IDS   Input Dependent Selector
LEDR   Level Encoded Dual-Rail
LETS   Level Encoded Transition Signaling
NCL   Null Convention Logic
NRZ   Non Return-to-Zero
OSS   Output Selection Stage
PPA   Parallel Prefix Adder
RZ   Return-to-Zero
SDS   Shift Dependent Selector
SI   Speed Independent
VLSI   Very-Large-Scale Integration

Chapter 1
Introduction to Asynchronous Circuit Design

1.1 Introduction

The digital circuit design style has evolved tremendously after the invention of transistors, and it can be categorised into synchronous and asynchronous design styles. The electronic devices developed immediately after the invention of transistors were designed without using a global clock and hence asynchronous in nature. Later, researchers have learned that having a global clock would allow for the development of smaller, faster circuits, leading to the emergence of synchronous design style. In fact, the synchronous paradigm was originally developed when resources were very expensive, and the cost of an individual computer element was very high. The designers wanted to maximise the efficiency of the hardware resources due to its high price; thus, the concept of a global clock signal was introduced to maximise the hardware reuse by scheduling their access and then ensuring that each step performs productive operation with each tick of the clock [1, 2].

The modern electronics industry is dominated by synchronous design style, with nearly all commercial products utilising completely synchronous circuits. This is because the synchronous design style is supported by a wide range of Electronic Design Automation (EDA) tools due to its ability to synchronise the circuit timing using one or more clock signals, resulting in compact and faster devices. The clock period in synchronous circuits is determined by the worst-case delay (refer Definition 2.7) of the circuit, which ensures that an ongoing computation is completed within a single clock cycle for all possible input–output combinations. The presence of a clock maintains synchronisation among all computing blocks, as the output of each computing block becomes valid and stable prior to the next clock period. The algorithms for any computation are organised in a manner such that it can fit in the clock paradigm and then implement them using synchronous clocks, where each clock tick results in a productive operation. The designers have certainly done an excellent job of mapping out, rewriting, and restructuring algorithms to fit this paradigm. However, since the time has evolved, the circuits are

P. Srivastava, *Completion Detection in Asynchronous Circuits*,
https://doi.org/10.1007/978-3-031-18397-3_1

designed to perform increasingly difficult tasks, and the majority of algorithms do not manifest themselves in this fashion. The circuit must be optimised to meet its timing specifications, a process referred to as timing closure [3]. As semiconductor miniaturisation technology advances, synchronous circuits encounter difficulties maintaining timing closure and synchronisation failures are caused due to clock skew. Another factor is the increased power consumption caused by the widespread distribution of high-frequency clock signals [4]. As a result, designers end up retrofitting all the logic, which is a standard practice in modern electronics [5].

Over the last few decades, the size of transistors has been reduced considerably due to advancement in VLSI technology, which results in higher switching speed of transistors. Furthermore, the reduced transistor size enables a single chip to accommodate a larger number of transistors, allowing it to execute more complicated and sophisticated operations. However, these technological advancements introduce new design challenges for circuit designers and the EDA tools that assist them. The continuous efforts to shrink transistor size to the atomic level in order to enhance circuit performance have resulted in transistors behaving less ideally. Moreover, scaling down the device size makes it difficult to leverage the high switching speed of transistors due to the constraints posed by clock skew and clock delay balancing. Several studies are ongoing to resolve these clock-related issues, and different design styles are being investigated, since the relative benefits of various circuit design styles keep changing as part of this constantly evolving technological environment. Designers are also experimenting with asynchronous design styles to address these timing issues while maintaining the performance boost of scaled devices, as discussed in the next section.

1.2 Evolution of Asynchronous Design Style

The evolution of asynchronous design can be divided roughly into four time zones. The fundamental theory of asynchronous circuit design was introduced between the 1950s and the early 1970s, proposed by Muller, Bartky [6] and Unger [7]. Asynchronous processors such as Atlas, Iliac, Iliac II, and MU-5 [8, 9] were also available commercially [10]. The next era, from the mid-1970s to the early 1980s, did not witness any significant development in the asynchronous design style, and the research work during this duration was more focused on the development of synchronous circuits. However, the asynchronous approach caught the attention of researchers once again between the mid-1980s and the late 1990s, and initial EDA tools to design asynchronous circuits were developed. A graph model known as STG (Signal Transition Graph) was used for an efficient synthesis of asynchronous control circuits [11]. Various companies and educational institutes like Philips Semiconductors, Myricom, and Caltech, University of Manchester, Tokyo Institute of Technology, etc. have adopted the asynchronous approach to improving the performance in terms of power consumption and modularity [10, 12].

The optimisation of the asynchronous approach still continues in the modern era from 2000 to present, to overcome the shortcomings of synchronous design due to the presence of a global clock. Asynchronous digital circuits are designed and commercialised by leading companies such as IBM, Intel, Philips Semiconductors, Sun Microsystems (now Oracle), etc. over the last few decades [13–17] with considerable cost benefits. Asynchronous circuit design has also been supported by a number of successful industrial experiments such as Intel RAPPID [18, 19], IBM FIR filter [20, 21], optimising continuous-time digital signal processors [22, 23], developing ultra-low-energy devices [24–27], system design to handle extreme temperature [28], and finally, developing alternative computing paradigms [29–31]; however, these experiments were not commercialised. One of the primary reasons for the less commercialisation of asynchronous circuits is the absence of sufficient mature asynchronous EDA tools [32, 33]. Several languages and design tools such as Tangram [17, 34–39], Communicating Hardware Processes or CHP [40–46], BALSA [47], asynchronous circuit compiler [48–51], Petrify [52–54], and various other tools [55–64] are being developed for asynchronous approach, as the asynchronous circuit design promises to overcome the limitations posed by the synchronous logic due to technology scaling [65].

1.3   Why Asynchronous?

Electronic circuit design has advanced tremendously since the transistor was invented, and it is common to design an electronic circuit with clock signal(s) to synchronise the computation process. However, the involvement of technology in daily life creates the need for fast and compact electronic devices with lower power consumption, hence imposing various constraints on the implementation of synchronous devices due to the presence of the global clock signal [66–68]. Clock skew control and delay balancing are difficult to manage due to technology scaling, as the clock signal must arrive at the same time to all sequential circuits [69–71]. Synchronous circuits devote 40% and more of their power to clock distribution [32, 72], and as designs become more complex, additional delay units are needed to manage the delay from the clock source to various flip-flops/latches in order to overcome clock skew [73, 74]. Moreover, synchronous circuits consider the worst-case delay to define the global clock time period, which limits the scope of increment in speed. Another key challenge is the demand for portable electronic devices with minimal power consumption [75, 76], while maintaining processing speed and silicon area [77]. The datapaths in synchronous circuits are always active, even if the data transmission line is idle, resulting in unnecessary power consumption.

Fig. 1.1 Asynchronous
design style

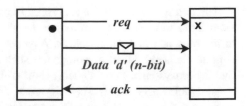

Asynchronous circuits can be used as an alternative to circumvent these challenges posed by clock, as clock signal(s) in such circuits are replaced by the *request*(*req*) and *acknowledge*(*ack*) signals, collectively known as handshaking signals.

Figure 1.1 illustrates how these handshaking signals are utilised to facilitate communication between two asynchronous logic blocks. The rectangular box with a dot (•) represents the module sending the data (or the sender), and the rectangular box with a cross (×) represents the module receiving the data (or the receiver), and the same notation is used throughout the book.

The asynchronous approach to designing a digital circuit is not new, and initially, processors were built using clockless techniques [78], such as IAS at Princeton [79] and ORDVAC at the University of Illinois [80]. Later, the development of asynchronous design style is overshadowed by the synchronous approach utilising a clock signal, as the presence of a global clock would simplify the implementation of the timing control due to its timing constraints. However, asynchronous design style has experienced a renaissance in recent decades, as the demand for smaller, faster circuits exacerbates the challenges associated with clock distribution, clock skew, and other clock-related issues in synchronous circuits. Figure 1.2 illustrates how an asynchronous design approach can be utilised to overcome challenges in meeting user requirements in a synchronous paradigm.

Asynchronous circuits are event-driven, which means that they can utilise the actual computation delay (refer Definition 2.9) corresponding to the event in process, unlike its synchronous equivalent that always waits for the worst-case delay for every computation. The term **event** is used in asynchronous design to refer to any circuit action that is initiated by a control signal or a data signal. The entire asynchronous paradigm essentially looks as follows: an asynchronous system consists of a large number of distinct components, all of which are waiting for a useful information to arrive. Since the system is not performing any operation, it usually runs at very low power. As soon as inputs arrive, the system activates the appropriate component, which performs useful computations and then produces

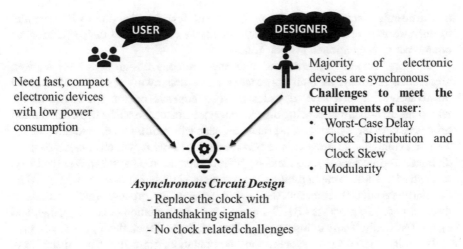

Need fast, compact electronic devices with low power consumption

Majority of electronic devices are synchronous
Challenges to meet the requirements of user:
- Worst-Case Delay
- Clock Distribution and Clock Skew
- Modularity

Asynchronous Circuit Design
- Replace the clock with handshaking signals
- No clock related challenges

Fig. 1.2 Why Asynchronous?

outputs that trigger further computation elsewhere in the system. The operation of the entire system is driven by the data moving around, and components are activated to perform useful computation only when they have useful information, which makes it all very efficient. The energy required to perform computations is not necessarily proportional to the number of components, but to the activity in the system; hence, all the active energy is used mostly for useful operations. Asynchronous approach is less susceptible to electromagnetic interference than synchronous design because asynchronous modules operate at different speeds and generate irregular current spikes. Moreover, since asynchronous circuits can utilise the actual computation delay corresponding to the event, the computation time can be reduced from worst-case delay to average-case delay (refer Definition 2.10) in such circuits.

1.3.1 Synchronous vs Asynchronous: How to Choose?

Asynchronous design style has several advantages over its synchronous counterpart, as discussed in the previous section. However, asynchronous approaches do have some drawbacks. Typically, implementing the handshaking in asynchronous circuits may result in a circuit speed, silicon area, and/or power consumption penalty. Therefore, it is crucial to analyse whether using the asynchronous approach to implement the circuit is worthwhile, i.e., whether using the asynchronous design style is providing a significant improvement in one or more of the aforementioned factors. In addition, asynchronous circuits are challenging to design due to the lack of readily available generic architecture, tools, languages, design constraints, etc. Nevertheless, asynchronous design style is used to implement several commercial and academic electronics circuits over the past few decades due to its potential to enhance the performance of digital circuits as compared to their synchronous counterparts, as mentioned in Sect. 1.2.

Choosing between synchronous and asynchronous design styles for a given application is tricky, as both approaches have their own set of advantages and disadvantages. The choice of a design style depends on the specifications for which the digital circuit is being developed. For instance, consider two computation processes with three distinct sub-processes, each with a different computation delay. The first computation process is represented by variable **A**, which consists of three distinct sub-processes A_1, A_2, and A_3 with computation delays δ_{A1}, δ_{A2}, and δ_{A3}, respectively. These computation delays are defined such that $\delta_{A1} \approx \delta_{A2} \approx \delta_{A3}$. Similarly, variable **B** represents the second computation process which consists of three distinct sub-processes B_1, B_2, and B_3 with computation delays δ_{B1}, δ_{B2}, and δ_{B3}, respectively. These computation delays are defined such that $\delta_{B1} > \delta_{B2} > \delta_{B3}$.

Now, in a clocked or synchronous computation, time has been discretised, resulting in discrete time windows that are evenly spaced by the clock period, i.e., the clock period is fixed. Figure 1.3 illustrates how to utilise the synchronous design style to implement the computation processes, **A** and **B**, each with its own distinct computation delay. Although the actual implementation is more complicated than that depicted in Fig. 1.3, it is explained here in simple terms to aid in comprehension of the design. A digital computation block (refer Definition 2.1) may perform multiple parallel computations, and these periodic boundaries are defined such that all parallel computations must be completed within the time constraints imposed by the periodic boundaries. The fixed periodic boundaries enable access to the final output data because the computation has been completed and the output is stable at the boundary, allowing a new computation process to begin at the next clock edge.

On the contrary, when a computation process is completed early in the asynchronous design style, other modules of the computation block that are waiting

Fig. 1.3 Synchronous design style

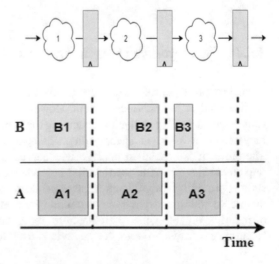

Fig. 1.4 Asynchronous
design style

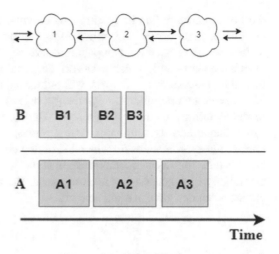

for the results can proceed as soon as the output data are stable and valid. The asynchronous implementation of the computation processes **A** and **B** is illustrated in Fig. 1.4. Although maintaining communication between different modules of the computation block incurs an overhead, the asynchronous paradigm eliminates the need to wait for the barrier to appear before initiating a new computation process. If this communication overhead in asynchronous paradigm is significantly less than the fixed clock period in the synchronous paradigm, then converting the synchronous computation block into an asynchronous one may result in a faster response.

It is evident from Figs. 1.3 and 1.4 that choosing between synchronous and asynchronous design styles depends upon the specifications of the application and its different submodules. If the computation process for an application is nicely balanced and the computation delays are perfectly uniform, then choosing synchronous design style would provide a better response than the asynchronous one, as the output would always be stable at the boundary defined by the fixed clock period. Considering the aforementioned computation processes, designing the computation process **A** using synchronous paradigm would be a wise choice as the computation delays are uniform, i.e., $\delta_{A1} \approx \delta_{A2} \approx \delta_{A3}$. Implementing process **A** using asynchronous paradigm incurs significant overhead and provides little benefit. On the other hand, if the computation process exhibits a wide range of computation delays, as in process **B**, then choosing asynchronous design style would provide a better response than the synchronous paradigm. The computation delays of the sub-processes are not uniform, i.e., $\delta_{B1} > \delta_{B2} > \delta_{B3}$, and hence the computation block implemented using asynchronous approach would operate with an average-case delay, resulting in a faster circuit.

Another reason to choose the asynchronous design approach is to minimise the power consumption in the circuit [5]. Synchronous systems are clocked systems that drain the power continuously. Additional measures must be taken to reduce the power consumption in synchronous circuits. On the other hand, asynchronous circuits become active only when useful information is available and go back to inactive state once the computation is done; hence the power consumption in asynchronous circuit is also event-driven, resulting in lower power consumption than its synchronous counterpart, without taking any additional measures [66, 81]. The silicon area required to implement a digital circuit is also a significant factor in determining the circuit design style. Hence, there is no clear winner between synchronous and asynchronous design styles; the choice is made based on the application specifications for which the digital circuit is being developed. A generic architecture for asynchronous design style is introduced in this book, as the availability of a generic architecture assists the researchers to make an informed choice when designing a digital circuit.

1.4   Book Organisation

Asynchronous design style does not rely on clock signal(s) to synchronise its events and thus is able to prevent clock-related challenges. Nevertheless, the majority of the researchers do not opt for asynchronous design style, and hence, it was vital to investigate why asynchronous design is not preferred, despite the various advantages it offers over its synchronous counterpart. The literature review unveils that asynchronous circuits encounter numerous challenges, as detailed in the following chapters. This book focuses on three primary aspects:

- The analysis of the existing circuit design styles and their timing constraints to determine the output data validity after completing the computation process. Various existing completion detection schemes are also examined in this analysis.
- The development of a generic deterministic completion detection scheme for asynchronous circuits using bundled data protocol.
- The development of a single-precision asynchronous bundled data barrel shifter to validate the generic design of the deterministic completion detection scheme. The generic design is also validated on a single-precision binary adder.

Chapter 2 provides the preliminaries required to conducting this research. Various technical terms used throughout the book are defined in this chapter. Chapter 3 highlights the importance of having an efficient completion detection scheme and reviews different existing completion detection schemes. In this review, a speculative completion detection scheme is examined that employs various delay models and an abort network. Furthermore, this chapter addresses the drawbacks of the speculative completion detection scheme and analyses a few existing works that implement the asynchronous adder with an early completion detection scheme without using an abort network. Chapter 4 covers the fundamentals of barrel shifter and binary adder, as the proposed generic architecture of deterministic completion detection scheme is validated on asynchronous bundled data barrel shifter and binary adder.

Chapter 5 outlines the design methodology that was employed to conduct this study. The chapter introduces a generic approach to determine the completion of an event deterministically for asynchronous circuits using bundled data protocol. In Chap. 6, a design example for implementing single-precision asynchronous bundled data barrel shifter and binary adder using the proposed generic approach is presented.

The conventional barrel shifter and the proposed asynchronous bundled data barrel shifter are both simulated using the Xilinx Vivado® Design Suite, and their results are compared in Chap. 7. The performance for a binary adder designed using the generic architecture is also compared with a clocked adder in this chapter.

1.5 Chapter Summary

This chapter provides an introduction to the asynchronous circuit design and discusses the advantages of replacing the global clock with handshaking signals. The background and evolution of the asynchronous design style are addressed. The chapter is concluded with a comparison of synchronous and asynchronous design styles. It is suggested that the optimal design style for an electronic circuit should be selected depending upon the application for which the circuit is being developed. The fundamental concepts of the asynchronous paradigm will be discussed in depth in the following chapters.

References

1. Manohar, R. (2015), 'Asynchronous circuits', IBM Research Cognitive Systems Colloquium: Brain-Inspired Computing at IBM Research - Almaden in San Jose, CA. [Cornell Tech].
2. Chu, P. P. (2006), RTL hardware design using VHDL: coding for efficiency, portability, and scalability, John Wiley & Sons.
3. Gimenez, G., Simatic, J. and Fesquet, L. (2019), From signal transition graphs to timing closure: Application to bundled-data circuits, in '2019 25th IEEE International Symposium on Asynchronous Circuits and Systems (ASYNC)', IEEE, pp. 86–95.
4. Semba, S. and Saito, H. (2021), 'Study on an RTL conversion method from pipelined synchronous RTL models into asynchronous RTL models'.
5. Manohar, R. (2020), 'Asynchronous logic: Design and EDA', UCSC Open Source Hardware and EDA Seminar. [Yale University].
6. Muller, D. and Bartky, W. (1959), 'A theory of asynchronous circuits, vol. xxix of the annals of the computation laboratory of Harvard university'.
7. Unger, S. H. (1973), 'Asynchronous sequential switching circuits', IEEE Transactions on Systems, Man, and Cybernetics SMC-3(3).
8. Goldstine, H. H. (1972), 'The computer from pascal to von Neumann', pp. 292–307.
9. Morris, D. and Ibbett, R. N. (1979), The architecture of the mu5 processor, in 'The MU5 Computer System', Springer, pp. 5–35.
10. Nowick, S. M. and Singh, M. (2015), 'Asynchronous design—part 1: Overview and recent advances', IEEE Design & Test 32(3), 5–18.

11. Chu, T.-A. (1987), Synthesis of self-timed VLSI circuits from graph-theoretic specifications, Massachusetts Institute of Technology.
12. Williams, T. E. and Horowitz, M. A. (1991), 'A zero-overhead self-timed 160-ns 54-b CMOS divider', IEEE Journal of Solid-State Circuits 26(11), 1651–1661.
13. Davies, M., Lines, A., Dama, J., Gravel, A., Southworth, R., Dimou, G. and Beerel, P. (2014), A 72-port 10g ethernet switch/router using quasi-delay-insensitive asynchronous design, in '2014 20th IEEE International Symposium on Asynchronous Circuits and Systems', IEEE, pp. 103–104.
14. Merolla, P. A., Arthur, J. V., Alvarez-Icaza, R., Cassidy, A. S., Sawada, J., Akopyan, F., Jackson, B. L., Imam, N., Guo, C., Nakamura, Y. et al. (2014), 'A million spiking neuron integrated circuit with a scalable communication network and interface', Science 345(6197), 668–673.
15. Fant, K. (2005), Logically determined design, Wiley Online Library.
16. Teifel, J. and Manohar, R. (2004), Highly pipelined asynchronous FPGAs, in 'Proceedings of the 2004 ACM/SIGDA 12th International Symposium on Field Programmable Gate Arrays', ACM, pp. 133–142.
17. Gageldonk, V. H., Kees, V. B., Peeters, A., Baumann, D., Gloor, D. and Stegmann, G. (1998), An asynchronous low-power 80c51 microcontroller, in 'Proceedings Fourth International Symposium on Advanced Research in Asynchronous Circuits and Systems', IEEE, pp. 96–107.
18. Parunak, H. V. D., Sauter, J., Fleischer, M. and Ward, A. (1999), 'The rapid project: Symbiosis between industrial requirements and mas research', Autonomous Agents and Multi-Agent Systems 2(2), 111–140.
19. Su, J. (2019), 'Full-swing dual-rail SRAM sense amplifier'. US Patent App. 15/839,375. Sutherland, I. E. (1989), 'Micropipelines', Communications of the ACM 32(6), 720–738.
20. Singh, M., Tierno, J. A., Rylyakov, A., Rylov, S. and Nowick, S. M. (2009), 'An adaptively pipelined mixed synchronous-asynchronous digital fir filter chip operating at 1.3 gigahertz', IEEE transactions on very large scale integration (VLSI) systems 18(7), 1043–1056.
21. Singh, M., Tierno, J. A., Rylyakov, A., Rylov, S. and Nowick, S. M. (2002), An adaptively pipelined mixed synchronous-asynchronous digital fir filter chip operating at 1.3 gigahertz, in 'Proceedings Eighth International Symposium on Asynchronous Circuits and Systems', IEEE, pp. 84–95.
22. Vezyrtzis, C., Jiang, W., Nowick, S. M. and Tsividis, Y. (2014), 'A flexible, event-driven digital filter with frequency response independent of input sample rate', IEEE Journal of Solid-State Circuits 49(10), 2292–2304.
23. Aeschlimann, F., Allier, E., Fesquet, L. and Renaudin, M. (2004), Asynchronous fir filters: towards a new digital processing chain, in '10th International Symposium on Asynchronous Circuits and Systems, 2004. Proceedings.', IEEE, pp. 198–206.
24. Chang, K.-L., Chang, J. S., Gwee, B.-H. and Chong, K.-S. (2013), 'Synchronous-logic and asynchronous-logic 8051 microcontroller cores for realizing the internet of things: A comparative study on dynamic voltage scaling and variation effects', IEEE journal on emerging and selected topics in circuits and systems 3(1), 23–34.
25. Christmann, J.-F., Beigne, E., Condemine, C., Leblond, N., Vivet, P., Waltisperger, G. and Willemin, J. (2010), Bringing robustness and power efficiency to autonomous energy harvesting microsystems, in '2010 IEEE Symposium on Asynchronous Circuits and Systems', IEEE, pp. 62–71.
26. Liu, T.-T., Alarcón, L. P., Pierson, M. D. and Rabaey, J. M. (2009), 'Asynchronous computing in sense amplifier-based pass transistor logic', IEEE transactions on very large scale integration (VLSI) systems 17(7), 883–892.
27. Nielsen, L. S. and Sparsø, J. (1999), 'Designing asynchronous circuits for low power: An IFIR filter bank for a digital hearing aid', Proceedings of the IEEE 87(2), 268–281.
28. Shepherd, P., Smith, S. C., Holmes, J., Francis, A. M., Chiolino, N. and Mantooth, H. A. (2013), A robust, wide-temperature data transmission system for space environments, in '2013 IEEE Aerospace Conference', IEEE, pp. 1–13.

29. Vacca, M., Graziano, M. and Zamboni, M. (2011), 'Asynchronous solutions for nanomagnetic logic circuits', ACM Journal on Emerging Technologies in Computing Systems (JETC) 7(4), 15.
30. Karaki, N., Nanmoto, T., Ebihara, H., Utsunomiya, S., Inoue, S. and Shimoda, T. (2005), A flexible 8b asynchronous microprocessor based on low-temperature poly-silicon TFT technology, in 'ISSCC. 2005 IEEE International Digest of Technical Papers. Solid-State Circuits Conference, 2005.', IEEE, pp. 272–598.
31. Peper, F., Lee, J., Adachi, S. and Mashiko, S. (2003), 'Laying out circuits on asynchronous cellular arrays: a step towards feasible nanocomputers?', Nanotechnology 14(4), 469.
32. Shin, Ziho et al. (2017). "Design of a clockless MSP430 core using mixed asynchronous design flow". In: IEICE Electronics Express, pp. 14–20170162.
33. Kondratyev, Alex and Kelvin Lwin (2002). "Design of asynchronous circuits using synchronous CAD tools". In: IEEE Design & Test of Computers 19.4, pp. 107–117.
34. Kessels, Joep, Torsten Kramer, Ad Peeters, et al. (2001). "DESCALE: a design experiment for a smart card application consuming low energy". In: European Low Power Initiative for Electronic System Design, pp. 247–262.
35. Kessels, Joep, Torsten Kramer, Gerrit Den Besten, et al. (2000). "Applying asynchronous circuits in contactless smart cards". In: Proceedings Sixth International Symposium on Advanced Research in Asynchronous Circuits and Systems (ASYNC 2000)(Cat. No. PR00586). IEEE, pp. 36–44.
36. Peeters, Ad and Kees van Berkel (1995). "Single-rail handshake circuits". In: Proceedings Second Working Conference on Asynchronous Design Methodologics. IEEE, pp. 53–62.
37. Van Berkel, Kees, Ronan Burgess, JoepKessels, et al. (1995). "A single-rail re-implementation of a DCC error detector using a generic standard-cell library". In: Proceedings Second Working Conference on Asynchronous Design Methodologies. IEEE, pp. 72–79.
38. Van Berkel, Kees, Ronan Burgess, Joep LW Kessels, et al. (1994). "A fully asynchronous low-power error corrector for the DCC player". In: IEEE Journal of Solid-State Circuits 29.12, pp. 1429–1439.
39. Van Berkel, Kees (1993). Handshake circuits: an asynchronous architecture for VLSI programming. Vol. 5. Cambridge University Press.
40. Martin, Alain J and Christopher D Moore (2011). "CHP and CHPsim: A language and simulator for fine-grain distributed computation". In: Dept. Comput. Sci., California Inst. Technol., Pasadena, CA, USA, Tech. Rep. CS-TR-1-2011.
41. Patel, Girish N et al. (2006). "An asynchronous architecture for modeling intersegmental neural communication". In: IEEE Transactions on Very Large Scale Integration (VLSI) Systems 14.2, pp. 97–110.
42. Boahen, Kwabena A (2004). "A burst-mode word-serial address-event link-I: Transmitter design". In: IEEE Transactions on Circuits and Systems I: Regular Papers 51.7, pp. 1269–1280.
43. Manohar, Rajit and Clinton Kelly (2001). "Network on a chip: modeling wireless networks with asynchronous VLSI". In: IEEE Communications Magazine 39.11, pp. 149–155.
44. Renaudin, Marc, P Vivet, and F Robin (1999). "ASPRO: an Asynchronous 16-bit RISC Microprocessor with DSP Capabilities". In: Proceedings of the 25th European Solid- State Circuits Conference. IEEE, pp. 428–431.
45. Martin, Alain J, Andrew Lines, et al. (1997). "The design of an asynchronous MIPS R3000 microprocessor". In: Proceedings Seventeenth Conference on Advanced Research in VLSI. IEEE, pp. 164–181.
46. Martin, Alain J, Steven MBurns, et al. (1989). The design of an asynchronous microprocessor. Tech. rep. California Inst of Tech Pasadena Dept of Computer Science.
47. Edwards, Doug and Andrew Bardsley (2002). "Balsa: An asynchronous hardware synthesis language". In: The Computer Journal 45.1, pp. 12–18.
48. Tiempo Secure TAM16: 16-Bit Microcontroller IP Core (n.d.). http://www.tiempoic.com/products/ip-cores/TAM16.html. Accessed: 2018-12-23.

49. Tiempo Secure ACC: Asynchronous Circuit Compiler (n.d.). http://www.tiempoic.com/products/sw-tools/acc.html. Accessed: 2018-12-23.
50. Yakovlev, Alex, Pascal Vivet, and Marc Renaudin (2013). "Advances in asynchronous logic: From principles to GALS & NoC, recent industry applications, and commercial CAD tools". In: Proceedings of the Conference on Design, Automation and Test in Europe. EDA Consortium, pp. 1715–1724.
51. Spear, Chris (2008). SystemVerilog for verification: a guide to learning the testbench language features. Springer Science & Business Media.
52. Akram, Tallha et al. (2017). "Towards real-time crops surveillance for disease classification: exploiting parallelism in computer vision". In: Computers & Electrical Engineering 59, pp. 15–26.
53. Cortadella, Jordi, Michael Kishinevsky, Alex Kondratyev, Luciano Lavagno, Enric Pastor, et al. (1999). "Petrify: a tool for synthesis of petri nets and asynchronous circuits". In: Software.
54. Cortadella, Jordi, Michael Kishinevsky, Alex Kondratyev, Luciano Lavagno, and Alexandre Yakovlev (1997). "Petrify: a tool for manipulating concurrent specifications and synthesis of asynchronous controllers". In: IEICE Transactions on information and Systems 80.3, pp. 315–325.
55. Handshake-Solution TideTimeless Design Environment (n.d.). http://www.handshakesolutions.com. Accessed: 2019-01-02.
56. DEMOSessionDEMOSession (n.d.). http://conferences.computer.org/async2007.Accessed: 2019-01-02.
57. Visual STG Lab S. F. a. H. Palbøl., Visual STG Lab (n.d.). http://vstgl.sourceforge.net/. Accessed: 2018-12-24.
58. NUS Asynchronous High Level Synthesis Tool (VERISYN) (n.d.). NUSEngineering. Accessed: 2018-12-24.
59. WorkCraft (n.d.). https://workcraft.org/. Accessed: 2018-12-24.
60. Bardsley, Andrew, Luis Tarazona, and Doug Edwards (2009). "Teak: A token-flow implementation for the balsa language". In: 2009 Ninth International Conference on Application of Concurrency to System Design. IEEE, pp. 23–31.
61. Bardsley, Andrew, Luis Tarazona, and Doug Edwards (2009). "Teak: Atokenflowimplementation for the balsa language". In: 2009 Ninth International Conference on Application of Concurrency to System Design. IEEE, pp. 23–31.
62. Kangsah, Ben et al. (2003). "DESI: A tool for decomposing signal transition graphs". In: 3rd ACiD-WG Workshop.
63. Endecott, Philip and Stephen B Furber (1998). "Modelling and Simulation of Asynchronous Systems Using the LARD Hardware Description Language." In: ESM, pp. 39–43.
64. Theodoropoulos, Georgios K, GK Tsakogiannis, and JV Woods (1997). "Occam: an asynchronous hardware description language?" In: EUROMICRO 97. Proceedings of the 23rd EUROMICRO Conference: New Frontiers of Information Technology (Cat. No. 97TB100167). IEEE, pp. 249–256.
65. Rahbaran, B. and Steininger, A. (2008), 'Is asynchronous logic more robust than synchronous logic?', IEEE Transactions on dependable and secure computing 6(4), 282–294.
66. Tabassam, Z., Naqvi, S. R., Akram, T., Alhussein, M., Aurangzeb, K. and Haider, S. A. (2019), 'Towards designing asynchronous microprocessors: From specification to tapeout', IEEE Access 7, 33978–34003.
67. Sadeghi, R. and Jahanirad, H. (2017), Performance-based clustering for asynchronous digital circuits, in '2017 Iranian Conference on Electrical Engineering (ICEE)', IEEE, pp. 238–243.
68. Srivastava, P., Chung, E. and Ozana, S. (2020), 'Asynchronous floating-point adders and communication protocols: A survey', Electronics 9(10), 1687.
69. Sparsø, J. (2020), Introduction to Asynchronous Circuit Design., DTU Compute, Technical University of Denmark.
70. Maitham Shams, J. C. E. and Elmasry, M. I. (2001), Asynchronous Circuits, Wiley Encyclopedia of Electrical and Electronics Engineering.

71. Renaudin, M. (2000), 'Asynchronous circuits and systems: a promising design alternative', Microelectronic engineering 54(1-2), 133–149.
72. Krstic, M., Grass, E. and Fan, X. (2017), Asynchronous and gals design-overview and perspectives, in '2017 New Generation of CAS (NGCAS)', IEEE, pp. 85–88.
73. Srivastava, N. and Manohar, R. (2018), 'Operation-dependent frequency scaling using desynchronization', IEEE Transactions on Very Large Scale Integration (VLSI) Systems 27(4), 799–809.
74. Donno, M., Ivaldi, A., Benini, L. and Macii, E. (2003), Clock-tree power optimization based on RTL clock-gating, in 'Proceedings of the 40th annual Design Automation Conference', ACM, pp. 622–627.
75. Wheeldon, A., Morris, J., Sokolov, D. and Yakovlev, A. (2019), 'Self-timed, minimum latency circuits for the internet of things', Integration 69, 138–146.
76. Oliveira, D. L., Verducci, O., Torres, V. L., Saotome, O., Moreno, R. L. and Brandolin, J. B. (2018), A novel architecture for implementation of quasi delay insensitive finite state machines, in '2018 IEEE XXV International Conference on Electronics, Electrical Engineering and Computing (INTERCON)', IEEE, pp. 1–4.
77. Leite, T. F. d. P. (2019), FD-SOI technology opportunities for more energy efficient asynchronous circuits, Grenoble Alpes.
78. Wilcox, S. P. (1999), Synthesis of asynchronous circuits, PhD thesis, University of Cambridge, Queens' College.
79. Burks, A. W., Goldstine, H. H. and Chaney, N. J. V. (1946), 'Preliminary discussion of the logical design of an electronic computing instrument', Report to US Army Ordnance Department.
80. ORDVAC Manual (1951), Ballistic Research Laboratories, University of Illinois. Palmer, J. F., Ravenel, B. W. and Nave, R. (1982), 'Numeric data processor'. US Patent 4,338,675.
81. Srivastava, P. and Chung, E. (2022), 'An asynchronous bundled-data barrel shifter design that incorporates a deterministic completion detection technique', IEEE Transactions on Circuits and Systems II: Express Briefs 69(3), 1667–1671.

Chapter 2
Preliminary Considerations for Asynchronous Circuit Design

2.1 Definitions

This section provides the definitions of several terms that are repeatedly used in this book. These definitions will aid in comprehension of the fundamental principle underlying asynchronous design styles and other related concepts.

Notation

ζ denotes a digital computation block with γ mutually exclusive signal paths. Each signal path has a different propagation delay based on the components present in the path. The set $\Delta = \{\delta_1, \delta_2, \ldots, \delta_\gamma\}$ represents the universal set of the propagation delays corresponding to the set $\mathcal{P} = \{\mathcal{P}_1, \mathcal{P}_2, \ldots, \mathcal{P}_\gamma\}$, where $\mathcal{P}$ denotes the universal set of mutually exclusive signal paths connecting inputs to outputs for ζ. The universal set of events for ζ is represented by $E = \{e_1, e_2, \ldots, e_\epsilon\}$, where ϵ denotes the total number of possible events, and the probability of occurring an event $e_i \in E$ is denoted by $prob(e_i)$. Variables $\epsilon, \gamma, i, j, k, n \in \mathbb{Z}_+$, where $\mathbb{Z}_+$ denotes the set of non-negative integers. Paths $\mathcal{P}_f$, $\mathcal{P}_t$, and $\mathcal{P}_c$ denote the false path, true path and critical path of ζ, respectively, and δ_{wc}, δ_{bc}, δ_{avg}, and δ_{ac} represent its worst-case delay, best-case delay, average-case delay, and actual computation delay, respectively.

Digital Computation Block

Definition 2.1 A digital computation block ζ is a module that operates exclusively on binary data. An n-bit wide ζ contains multiple signal paths connecting inputs to output, with the total number of mutually exclusive signal paths denoted by γ. The active signal path associated with an ongoing event is determined by the properties of that event.

© The Author(s), under exclusive license to Springer Nature Switzerland AG 2022
P. Srivastava, *Completion Detection in Asynchronous Circuits*,
https://doi.org/10.1007/978-3-031-18397-3_2

Cardinality

Definition 2.2 The cardinality of a set denotes the number of elements present in the set. Hence,
cardinality of the universal set of events $|E| = \epsilon$,
cardinality of the universal set of signal paths $|\mathcal{P}| = \gamma$, and
cardinality of the universal set of computation delay $|\Delta| = \gamma$.

Signal Path

Definition 2.3 A signal path in a digital circuit is defined as either a wire connecting two logic gates or a path within a logic gate from one of its input ports to its output port.

False Path

Definition 2.4 For any digital computation block ζ, $\exists\, \mathcal{P}_f \in \mathcal{P} : \mathcal{P}_f$ is never activated $\forall\, e_i \in E$. In other words, a false path is a signal path in the circuit that is not used to propagate a signal. The term *'false path'* is used in this book to describe a signal path of the computation block ζ that cannot be activated by any combination of inputs. False paths are common in electronic circuits, and they exist due to the design of the circuit or the way it is used.

True Path

Definition 2.5 For any computation block ζ, $\exists\, \mathcal{P}_t \in \mathcal{P} : \mathcal{P}_t$ is activated at least for one event $e_i \in E$. In other words, a true path is defined as a signal path connecting inputs to outputs of a computation block and is not a false path.

Critical Path

Definition 2.6 For any digital computation block ζ, $\exists \mathcal{P}_c \in \mathcal{P}$: its computation delay $\delta_{wc} = \max \{\delta_1, \delta_2, ..., \delta_\gamma\} \ \forall \ e_i \in E$. In other words, the critical path in a circuit is defined as the signal path connecting inputs to outputs with the largest delay, for all possible events from the universal set E.

Worst-Case Delay

Definition 2.7 For any digital computation block ζ, $\exists \ \delta_{wc} = \max \{\delta_1, \delta_2, ..., \delta_\gamma\}$: $\delta_i \leq \delta_{wc} \ \forall \ \delta_i \in \ \Delta$ and $e_i \in E$. In other words, the term worst-case delay refers to the computation delay associated with the critical path of a digital circuit for all possible events from the universal set E. The time period of the clock signal in synchronous design style is determined using the worst-case delay, and it is always greater than the worst-case delay in order to provide an error-free and valid output.

Best-Case Delay

Definition 2.8 For any computation block ζ, $\exists \ \delta_{bc} = \min \{\delta_1, \delta_2, ..., \delta_\gamma\}$: $\delta_i \geq \delta_{bc}$ $\forall \ \delta_i \in \ \Delta$. In other words, the best-case delay of a computation block is defined as the smallest delay associated with the signal path connecting inputs to outputs, for all possible events from the universal set E.

Actual Computation Delay

Definition 2.9 For any digital computation block ζ, the actual computation delay $\delta_{ac} \in \delta$ refers to the delay associated with the active signal path $\mathcal{P}_{ac} \in \mathcal{P}$ required to process an event $e_i \in E$. The event-driven nature of asynchronous design style allows the circuit to indicate the completion of an event after actual computation delay. The value of $\mathcal{P}_{ac}$ lies between the best-case and worst-case delays of ζ i.e., $\mathcal{P}_{bc} \leq \mathcal{P}_{ac} \leq \mathcal{P}_{wc}$.

Average-Case Delay

Definition 2.10 The average-case delay is the mean time needed by the digital computation block ζ to perform operations for all possible input–output combinations. The average-case delay can be calculated as

$$\delta_{avg} = \sum_{i=1}^{\gamma} \left(\frac{\delta_i . r_i}{\epsilon} \right), \tag{2.1}$$

where r_i represents the repetitivity of the signal path $\mathcal{P}_i$ with delay δ_i, for all possible events (ϵ) from the universal set E. In other words, r_i can be defined as the number of times path $\mathcal{P}_i$ is repeated by events $e_i \in E_k \subset E : E_k = \{e_i \mid e_i$ needs the computation delay $\delta_i \in \Delta$ and $e_i \notin E - E_k\}$. The term (r_i / ϵ) gives the probability of event e_i, and Eq. 2.1 can be rewritten as

$$\delta_{avg} = \sum_{i=1}^{\gamma} (\delta_i . prob(e_i)). \tag{2.2}$$

The computation delay of an asynchronous circuit can be determined by an average-case delay, as the asynchronous design style utilises the actual computation delay of the ongoing event.

2.2 Delay Assumptions

Once the fundamentals of a digital circuit are defined, this chapter would proceed to discuss the presumptions made for various circuit delays. The initial research work on asynchronous systems started with the assumption of gate and wire delays present in digital circuits. An asynchronous system with different gate and wire delays are shown in Fig. 2.1. These delays can be assumed bounded or unbounded, and asynchronous circuits can be distinguished based on the delay assumptions utilised, as shown in Table 2.1 and discussed in the next few sections.

In_1, In_2, In_3, In_4: Input Signals
α, β, γ: Intermediate Signals
Out_1, Out_2: Output Signals
A, B, C: Gates
$\delta_A, \delta_B, \delta_C$: Gate Delays
w_1, w_2, w_3: Wire Delays

Fig. 2.1 Delay assumptions

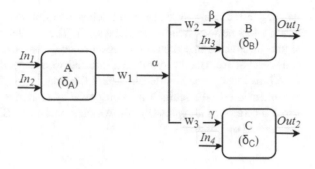

Table 2.1 Delay assumptions

	Gate delays			Wire delays		
Delay model	δ_A	δ_B	δ_C	w_1	w_2	w_3
Fundamental mode	Bounded delay			Bounded delay		
Speed independent mode	Unbounded delay			0	0	0
Delay-insensitive mode	Unbounded delay			Unbounded delay		
Quasi-delay-insensitive mode	Unbounded delay			w_1	$(w_2 \approx w_3) < (\delta_B \approx \delta_C)$	

2.2.1   Fundamental Mode or Huffman Circuit

In the fundamental mode, the gate and wire delays are assumed to be bounded [1], i.e., the upper and lower bounds of the gate and wire delays are known [2]. These upper and lower bounds t_1 and t_2, respectively, ensure the stability of asynchronous circuits. Two input transitions that occurred in less than t_1 time are considered as a single input, and two transitions needing more than t_2 time are considered as two sequential inputs [3]. However, these circuits are prone to hazards if the signal changes between the specified bound t_1 and t_2. Therefore, some assumptions are made to avoid hazards: either add internal delay models with an accurate value of the computation time required by the circuit or restrict to change only one input signal at a time. The bounded delay approach to determine process completion is slower, its design is complex, and it is difficult to detect the fault in the circuit [4].

2.2.2   Speed Independent or Muller Circuit

The bounded delay assumption used by fundamental mode restricts the performance of asynchronous circuits due to the action taken for hazard removal. David Muller has suggested using an unbounded delay assumption for gates, whereas wires are assumed to be ideal with zero delays [5]. The Muller circuit sends a completion signal to indicate that the computation is done, and the circuit is ready to receive a new input [6]. Such circuits are referred to as Speed Independent (SI) circuits [7, 8], as the gate delays do not affect the final state of the asynchronous circuit. An extension of SI circuits is also proposed by assuming the presence of zero-delay input inverters on all gates [9, 10], which is not a true SI circuit, but it can provide the desired outcome.

2.2.3   Delay-Insensitive Circuit

The SI asynchronous circuits assumed zero wire delays, but as the circuit size is continuously being scaled-down, the wire delays dominate the gate delays [11]. The delay-insensitive (DI) circuit is introduced by the Macromodule computer systems [12, 13]. The DI asynchronous circuits assume a finite unbounded gate and wire delay, which is an extremely robust delay assumption. However, such circuits are not very practical, and DI behaviour is restricted to the circuits having only inverter and Muller C-element [14]. C-element is a standard logic gate commonly used to synchronise independent processes in most asynchronous electronic circuits [15]. Researchers have proposed various modules and algorithms to implement more flexible DI circuits [16–18].

2.2.4   Quasi-Delay-Insensitive Circuit

The fact that DI asynchronous circuits can be implemented by using only C-elements and inverters restricts the flexibility to design most of the DI circuits. *Isochronic forks* are introduced by modifying DI asynchronous circuits, which allows matching the delay in certain conditions [19]. The system shown in Fig. 2.1 will act as a Quasi Delay Insensitive (QDI) circuit when the wire delays $w_2 = w_3$ or more accurately, $(w_2 \simeq w_2) < (\delta_A \simeq \delta_B)$. This assumption allows to implement a wide variety of QDI asynchronous circuits, but it should be designed carefully as it might not provide the desired response for some applications [20, 21].

2.2.5   Other Approaches

The delay assumptions are further relaxed to improve the performance of QDI circuits, by making a riskier delay assumption. This arrangement is known as extended isochronic forks or Quasi QDI [22]. *Field forks* are introduced to improve the performance at the transistor level [23]. Another approach by utilising bounded wire delays and unbounded gate delays was also introduced [24]. Later, several simple and complex delay assumptions are used: unbounded complex-gate assumptions [25], bounded complex-gate assumptions [26], and bounded simple-gate assumptions [27]. All the delay assumptions for various asynchronous circuits discussed above have their own advantages and disadvantages, and the selection of the best suitable delay model depends upon the application.

2.3 Communication Protocols

Asynchronous circuits do not have a global clock; hence detecting the completion of the computational process has always been a challenge for the asynchronous design approach, and a communication protocol is needed to generate an *ack* signal after the computation is done [28–31]. The event-driven nature of an asynchronous circuit allows the circuit to become active only after receiving a request signal *req* and returns back to an inactive state after the computation is done, which is indicated by asserting the acknowledgment signal *ack* [32, 33]. A communication protocol is used to control this process in digital asynchronous circuits, as discussed in the next few subsections.

2.3.1 Data Encoding

The data encoding technique assists the asynchronous circuit to ensure the completion of an event before asserting the acknowledge signal, and the following are the most widely accepted encoding protocols:

1. **Bundled Data Scheme:** Bundled data schemes, like synchronous circuits [34], are often known as single-rail protocols since they use only one wire to encode a single bit of data *d* [35]. The circuit utilises the *req-ack* pair, as shown in Fig. 2.2; the computation starts after receiving a *req* signal, and the *ack* signal is asserted after waiting for the worst-case delay determined by the circuit's critical path for all input combinations, to ensure that the computation is done and the output data are valid.

Fig. 2.2 Bundled data protocol

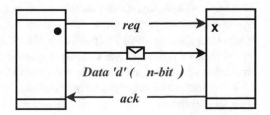

Fig. 2.3 Dual-Rail protocol

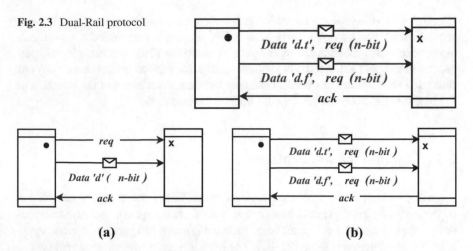

Fig. 2.4 Push channels for (**a**) bundled data and (**b**) dual-rail protocol

2. **Dual-Rail Protocol:** In dual-rail protocol, an n-bit data d is encoded using $2n$ wires, and signals $d.t.$ and $d.f.$ represent the true and false value of data, respectively, as shown in Fig. 2.3. The req signal is encoded within the data itself, and one extra wire is used to send the ack signal [36, 37].

   This protocol contains the information of the completion of an event within the data, and therefore the worst-case waiting period is required only when the critical path is used by the circuit, unlike bundled data protocol where the worst-case delay is used by default for all the cases. Once the data have arrived at the receiver, the value of data d can be determined by observing the $d.t.$ and $d.f.$ signals with the help of a completion detection circuit, which assists to generate the ack signal right after the computation is done and the output data are valid. However, the completion detection scheme begins to detect the valid data from $d.t.$ and $d.f.$ signals after the computation is done, causing an additional delay and therefore failing to produce the expected outcome for several applications.

2.3.2 Data Flow Direction

The req and ack signals can decide the direction of data flow on a channel. The channel where request can be initiated by the sender and the receiver sends the acknowledgment is known as push channel, as shown in Fig. 2.4.

   Alternatively, the receiver can also request data from the sender, and this kind of channel is known as pull channel, as shown in Fig. 2.5.

   The direction of request and acknowledge signals will be reversed for the pull channel and the data validity information will be encoded in the acknowledgment signal.

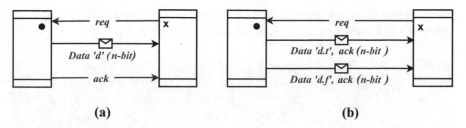

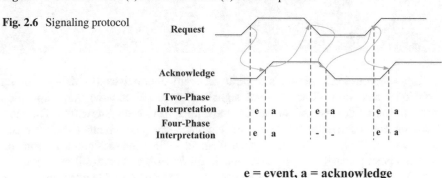

Fig. 2.5 Pull channels for (**a**) bundled data and (**b**) dual-rail protocol

Fig. 2.6 Signaling protocol

2.3.3 Signaling Protocols

The handshaking signals req and ack can be organised by using either two phase or four phase, as shown in Fig. 2.6 [38]. In the four-phase signaling protocol, the rising edge is used to encode the request and acknowledge signals, and the falling edge does not have any meaning. It is also known as RZ or level signaling. It consumes unnecessary time and energy because of its return-to-zero transition. On the contrary, the two-phase protocol encodes the handshaking signals to both rising and falling edges; therefore, it is also known as NRZ or transition signaling.

Ideally, the two-phase protocol should result in faster and energy-efficient designs. Sutherland encouraged the transition signaling and proposed a two-phase bundled data protocol for the first time [39], and an asynchronous design for ARM microprocessors was developed [40], inspired by the framework proposed by Sutherland. Later, the Amulet team found that two-phase latches consume more area, as the implementation of the two-phase protocol is complex and requires more logic for implementation [38]. It can be a possibility that a two-phase protocol may require more power due to increased complexity compared to the power consumption spared by the reduced transitions. Davis and Nowick [41] assessed different design styles and proposed a four-phase pipeline design, which is energy-efficient, smaller, and faster than a two-phase design.

Since the choice of a communication protocol is dependent on the specifications of the application for which the asynchronous circuit is being developed, it can

be concluded from the preceding discussion that a single communication protocol
suitable for all asynchronous applications cannot be defined.

> The communication protocol used to control the processing of events can be
> expressed as the cross product of the various data related information as given
> below:
>
> (Data Encoding Technique) $\times$ (Data Flow Direction) $\times$ (Signaling Protocol).
>
> $$(2.3)$$

The solution space contains several different alternatives for data encoding
technique, data flow direction, and signaling protocol, and the communication
protocol can be defined by combining one choice from each alternative. The most
common communication protocols are bundled data push channel with 2-phase
signaling protocol, bundled data push channel with 4-phase signaling protocol,
dual-rail push channel with 4-phase signaling protocol, and dual-rail push channel
with 2-phase signaling protocol; however, there are other possible alternatives. The
choice of protocol would affect the performance of circuit implementation in terms
of area, power consumption, speed, robustness, etc. This book focuses on the circuits
that utilise bundled data push channel with 4-phase signaling protocol and assume
that the sender initiates the request, unless stated otherwise.

2.4 Necessity of Completion Detection in Asynchronous Design

Asynchronous circuits differ significantly from synchronous circuits in a number
of ways, and it is intriguing how these distinctions could be utilised to address
the challenges that synchronous design style poses on modern electronic circuits.
The possibility that asynchronous design style could outperform synchronous ones
in digital circuit performance has encouraged several researchers and industries
to implement asynchronous devices, as discussed in Sect. 1.2. The absence of
clock signal(s) in asynchronous circuits eliminates the clock-related complexity and
facilitates design modularity. The computation delay of an asynchronous circuit is
determined by the actual input data, and it can be built as an event-driven model
despite using the worst-case delay model employed by synchronous circuits.

The event-driven model of asynchronous design style can assist the system to sequence its computational process, i.e., the system knows **where** it needs to communicate **what** information, but it cannot tell **when** a component's computation is complete. Hence, asynchronous circuits need a completion detection scheme to detect **when** a component's computation is complete [42], and most completion detection systems fall into one of the two categories, as discussed in Sect. 2.3.1: (i) bundled data protocol and (ii) dual-rail protocol.

It is evident from the above discussion in Sect. 2.3 that there is no optimal choice of a communication protocol for asynchronous circuits. When compared to the synchronous design, the dual-rail protocol uses two wires per bit, resulting in greater silicon area and power consumption. Moreover, a completion detection network is required to detect data validity **after** the computation is done, which adds extra gate delay to the circuit computation delay. Bundled data protocol uses single-rail implementation, and therefore power consumption and silicon area are comparable to the synchronous design; yet, it cannot take the advantage of the event-driven attribute of asynchronous circuits as it utilises the worst-case delay to indicate a valid output regardless of actual input data.

Nowick proposed a completion detection scheme with multiple delay models for bundled data asynchronous circuits [43]. This completion detection scheme is designed for asynchronous adders to speculate the circuit delay depending upon input data value, and a delay model is selected based on that calculation. However, if the circuit delay associated with the input data exceeds the selected model delay, an abort network is utilised to abort the selected speculative path, and the path corresponding to the worst-case delay model is chosen by default. This technique utilises single-rail implementation, making silicon area comparable to the synchronous design and provides an average-case delay depending upon actual input data. Unlike dual-rail coding, which begins detecting data validity after the computation is completed, the completion detection network in this approach operates in parallel with the computation and has reported improved processing speed ranges from 5% to 30% over a comparable synchronous implementation. However, the performance of this approach is highly dependent upon the efficiency of the abort network, as it should abort the selected speculative path in case of false triggering of ack signal due to incorrect delay estimation.

The major drawback of the speculative approach is that not only the logic for late completion detection is speculative, but it also introduces additional delay and logic depth to the adder implementation. Lai observed that under certain conditions, Nowick's speculative computation may fail to abort an early completion, resulting in an incorrect outcome [44]. In order to indicate the completion of an event without using the abort network, Lai modified Nowick's concept and proposed an early completion detection scheme for asynchronous adders [44, 45]. Lai's work was limited to asynchronous adders only, but it opens the door for a design scheme for asynchronous circuits using bundled data protocol, which can determine the completion of an event and data validity deterministically. The design scheme adopts single-rail coding, and the completion detection circuit operates in parallel with the primary computation process, resulting in a computation delay and silicon area comparable to that of the synchronous counterpart.

2.5 Chapter Summary

This chapter discusses the fundamentals of asynchronous design style that must be known prior to implementing asynchronous circuits. Different terminologies, notations, and definitions used throughout this book are defined in this chapter. The categorisation of asynchronous circuits is also discussed based on the delay assumptions and communication protocols used to implement the circuit. Since the clock signal is absent in asynchronous circuits, a completion detection scheme is required to indicate the data validity. Various existing completion detection schemes for asynchronous design style are examined in the next chapter.

References

1. Huffman, David A (1954a). "The synthesis of sequential switching circuits". In: Journal of the franklin Institute 257.3, pp. 161–190.
2. Huffman, David A (1954b). "The synthesis of sequential switching circuits". In: Journal of the franklin Institute 257.3, pp. 161–190.
3. Unger, Stephon H (1993). "A building block approach to unlocked systems". In: [1993] Proceedings of the Twenty-sixth Hawaii International Conference on System Sciences. Vol. 1. IEEE, pp. 339–348.
4. Hauck, Scott (1995). "Asynchronous design methodologies: An overview". In: Proceedings of the IEEE 83.1, pp. 69–93.
5. Muller, David E (1956). "A theory of asynchronous circuits". In: Report 75, University of Illinois.
6. Muller, David E (1963). "Asynchronous logics and application to information processing". In: Switching Theory in Space Technology, pp. 289–297.
7. Miller, Raymond Edward (1965). "Switching theory. Volume 2- Sequential circuits and machines(Book on unified treatment of switching theory, emphasizing synthesis and analysis aspects of switching circuits)". In: NEW YORK, JOHN WILEY AND SONS, INC., 1965. 250 P.
8. Muller, DE and WS Bartky (1957). "A theory of asynchronous circuits II". In: Digital Computer Laboratory 78.
9. Kondratyev, Alex et al. (1994). "Basic gate implementation of speed-independent circuits". In: 31st Design Automation Conference. IEEE, pp. 56–62.
10. Beerel, Peter A and Teresa HY Meng (1992). "Automatic gate-level synthesis of speed independent circuits". In: ICCAD, pp. 581–586.
11. Compass Design Automation Inc. (1995). "Deep submicron seminar".
12. Clark, Wes A and Charles E Molnar (1974). "Macromodular computer systems". In: Computers in Biomedical Research 4, pp. 45–85.

13. Clark, Wesley A (1967). "Macromodular computer systems". In: Proceedings of the April 18–20, 1967, spring joint computer conference. ACM, pp. 335–336.
14. Martin, Alain J (1999b). "The limitations to delay-insensitivity in asynchronous circuits". In: Beauty is our business. Springer, pp. 302–311.
15. Nguyen, Nam-phuong et al. (2010). "Design and analysis of a robust genetic Muller C-element". In: Journal of theoretical biology 264.2, pp. 174–187.
16. Patra, Priyadarsan, Stanislav Polonsky, and Donald S Fussell (1997). "Delay insensitive logic for RSFQ superconductor technology". In: Proceedings Third International Symposium on Advanced Research in Asynchronous Circuits and Systems. IEEE, pp. 42–53.
17. Patra, Priyadarsan and Donald S Fussell (1994). "Efficient building blocks for delay insensitive circuits". In: Proceedings of 1994 IEEE Symposium on Advanced Research in Asynchronous Circuits and Systems. IEEE, pp. 196–205.
18. Keller, Robert M (1974). "Towards a theory of universal speed-independent modules". In: IEEE Transactions on Computers 100.1, pp. 21–33.
19. Martin, Alain J (1990a). "The design of a delay-insensitive microprocessor: An example of circuit synthesis by program transformation". In: Hardware Specification, Verification and Synthesis: Mathematical Aspects. Springer, pp. 244–259.
20. Nanya, Takashi et al. (1994). "TITAC: Design of a quasi-delay-insensitive microprocessor". In: IEEE Design & Test of computers 11.2, pp. 50–63.
21. Van Berkel, Kees (1992). "Beware the isochronic fork". In: Integration, the VLSI journal 13.2, pp. 103–128.
22. Berkel,Kees van, Ferry Huberts, and Ad Peeters (1995). "Stretching quasi delay insensitivity by means of extended isochronic forks". In: Proceedings Second Working Conference on Asynchronous Design Methodologies. IEEE, pp. 99–106.
23. Kishinevsky, Michael et al. (1994). Concurrent hardware: the theory and practice of self-timed design. John Wiley & Sons, Inc.
24. Armstrong, Douglas B, Arthur D Friedman, and Premachandran R Menon (1969). "Design of asynchronous circuits assuming unbounded gate delays". In: IEEE Transactions on Computers 100.12, pp. 1110–1120.
25. Chu, Tam-Anh (1987). "Synthesis of self-timed VLSI circuits from graph-theoretic specifications". PhD thesis. Massachusetts Institute of Technology.
26. Moon, Cho W, Paul R Stephan, and Robert K Brayton (1991). "Synthesis of hazard-free asynchronous circuits from graphical specifications". In: 1991 IEEE International Conference on Computer-Aided Design Digest of Technical Papers. IEEE, pp. 322–325.
27. Lavagno, Luciano, Kurt Keutzer, and Alberto L Sangiovanni-Vincentelli (1995). "Synthesis of hazard-free asynchronous circuits with bounded wire delays". In: IEEE Transactions on Computer-Aided Design of Integrated Circuits and Systems 14.1, pp. 61–86.
28. Sparsø, J. (2020), Introduction to Asynchronous Circuit Design., DTU Compute, Technical University of Denmark.
29. Nowick, S. M. and Singh, M. (2015), 'Asynchronous design—part 1: Overview and recent advances', IEEE Design & Test 32(3), 5–18.
30. Delvai, M. and Steininger, A. (2006), Solving the fundamental problem of digital design—a systematic review of design methods, in '9th EUROMICRO Conference on Digital System Design (DSD'06)', IEEE, pp. 131–138.
31. Sparsø, J. and Furber, S. (2002), Principles asynchronous circuit design, Springer. SPEC Benchmark Release 2/92 (1992).
32. Srivastava, P., Chung, E. and Ozana, S. (2020), 'Asynchronous floating-point adders and communication protocols: A survey', Electronics 9(10), 1687.
33. Srivastava, P. and Chung, E. (2022), 'An asynchronous bundled-data barrel shifter design that incorporates a deterministic completion detection technique', IEEE Transactions on Circuits and Systems II: Express Briefs 69(3), 1667–1671.
34. Peeters, A. and van Berkel, K. (1995), Single-rail handshake circuits, in 'Proceedings Second Working Conference on Asynchronous Design Methodologies', IEEE, pp. 53–62.

35. Peeters, Ad and Kees van Berkel (1995). "Single-rail handshake circuits". In: Proceedings Second Working Conference on Asynchronous Design Methodologies. IEEE, pp. 53–62.
36. Toosizadeh, N. (2010), Enhanced synchronous design using asynchronous techniques, PhD thesis, University of Toronto.
37. Gilchrist, B., Pomerene, J. H. and Wong, S. (1955), 'Fast carry logic for digital computers', IRE Transactions on Electronic Computers 4, 133–136.
38. Wilcox, S. P. (1999), Synthesis of asynchronous circuits, PhD thesis, University of Cambridge, Queens' College.
39. Sutherland, I. E. (1989), 'Micropipelines', Communications of the ACM 32(6), 720–738.
40. Furber, S. (1995), Computing without clocks: Micropipelining the arm processor, in 'Asynchronous Digital Circuit Design', Springer, pp. 211–262.
41. Davis, A. and Nowick, S. M. (1997), 'An introduction to asynchronous circuit design', The Encyclopedia of Computer Science and Technology 38, 1–58.
42. Manohar, R. (2020), 'Asynchronous logic: Design and EDA', UCSC Open Source Hardware and EDA Seminar. [Yale University].
43. Nowick, S. M. (1996), 'Design of a low-latency asynchronous adder using speculative completion', IEE Proceedings-Computers and Digital Techniques 143(5), 301–307.
44. Lai, K. K. (2016), Novel Asynchronous Completion Detection For Arithmetic Datapaths, Taylor's University, Malaysia.
45. Lai, K. K., Chung, E. C., Lu, S.-L. L. and Quigley, S. F. (2014), 'Design of a low latency asynchronous adder using early completion detection', Journal of Engineering Science and Technology 9(6), 755–772.

Chapter 3
Completion Detection Schemes for Asynchronous Design Style

3.1 Introduction

A completion detection circuit is the key element of asynchronous circuits, as it helps to determine the data validity after performing a computation. Asynchronous circuits do not have a global clock to indicate the process completion, and handshaking signals are used to communicate between two asynchronous logic blocks. The use of handshaking signals provides an opportunity to design asynchronous circuits such that the process completion can be determined depending upon actual input data, rather than waiting for critical path delay for every computation in case of a global clock. However, the absence of a global clock in asynchronous circuits creates the need for a completion detection technique to indicate that a process is done, and the data is valid. As asynchronous circuits are event-driven, a circuit becomes active only after receiving a request signal and goes back to an inactive state once the process is complete and sends the acknowledge signal. This process is controlled by the communication protocol in asynchronous circuits with the help of data encoding, as detailed in the next few sections of this chapter.

3.2 M-of-N Encoding Protocol

One of the preferred approaches to implement the completion detection circuit is by using M-of-N encoding, in which M inputs are set to logic 1 for an N-bit code, and $log_2 N$ bits can be represented using N wires with one extra wire used to send the ack signal [1]. The circuits with such encoding are known as DI circuits as it assumes a finite gate and wire delays [2–4]. Dual-rail coding is also a special case of M-of-N encoding with $M = 1$ and $N = 2$ [5]. Level Encoded Dual-Rail (LEDR) protocol utilises dual-rail coding, in which the code contains a 1-bit phase along with the data, and the phase keeps alternating between even and odd for every code

© The Author(s), under exclusive license to Springer Nature Switzerland AG 2022
P. Srivastava, *Completion Detection in Asynchronous Circuits*,
https://doi.org/10.1007/978-3-031-18397-3_3

Table 3.1 Delay Assumptions [8]	Message	1-of-4 Code	Dual-Rail code			
			True 1	False 1	True 0	False 0
	00	0001	0	1	0	1
	01	0010	0	1	1	0
	10	0100	1	0	0	1
	11	1000	1	0	1	0

such that two consecutive codes always have different phases [6]. A Level Encoded Transition Signaling (LETS) protocol is introduced by [7], which can be said as the generalisation of LEDR protocol. LETS does not require RZ phase, and only one rail switch per data transaction is needed, resulting in power and throughput advantages.

Another variant of 1-of-N encoding is for $N = 4$, also known as one hot encoding, in which n-bit data is represented by $2n$ wires. It is different from dual-rail encoding, as dual-rail coding encodes each bit using two wires, whereas one hot encoding encodes two-bit data using a four-bit unique code, as shown in Table 3.1 [8]. Even though both the coding schemes need equivalent silicon area, the one hot encoding is more energy efficient due to the fewer transitions. Null Convention Logic (NCL) is another example of 1-of-2 encoding with four-phase protocol, and the return-to-zero phase is used to separate two codes [9–11].

The circuits utilising M-of-N encoding are event-driven and can indicate the data validity as soon as the computation is done. However, this approach still needs a completion detection circuit to select the valid data available at the output. Asynchronous circuits with M-of-N encoding utilise Muller's C-element to indicate the completion of an event, as discussed in the following section.

3.2.1 Muller's C-Element

A digital circuit can be defined as a set of logic gates, and every change to the output of a single logic gate affects the outputs of the remaining logic gates. The circuit will generate the final output after the outputs of all the intermediate logic gates are stable and valid. Since the propagation delay of the various logic gates in a circuit varies, the circuit may contain intermediate states and multiple signal transitions before generating the final output.

As mentioned before, synchronous circuits utilise one or more clock signals to define their timing constraints, and the time period of the clock signal is determined by the worst-case delay of circuit's critical path for all possible events. Therefore, multiple signal transitions do not affect the final output of synchronous circuits, as the output will always be stable prior to the arrival of the next clock pulse. On the contrary, every signal transition in asynchronous circuits corresponds to a unique data, and hence multiple signal transitions should be avoided. Moreover, certain applications need the data to remain valid at all times [12], since there is no means to detect the completion of an event. Therefore, such circuits require a completion detection scheme to assist them in asserting the *ack* signal by detecting the valid output.

Example

A two-input AND gate outputs logic 1 only when both inputs are set to logic 1. However, if the output of the AND gate is logic 0, it indicates that either both or one of the inputs has been set to logic 0, but it is difficult to predict which one. The same is true for the OR gate, which produces a logic 0 output only when both inputs are set to logic 0, otherwise, it provides a logic 1 output, but it cannot detect which input/s are changed.

David Muller invented the C-element [13], which functions as a state-holding component, similar to an asynchronous set–reset latch. The output of a two-input C-element changes only when both the inputs have same values, otherwise it will store its previous state and the output remains unchanged. In other words, $i_1 = i_2 = 0 \implies out = 0$; $i_1 = i_2 = 1 \implies out = 1$; $(i_1 = 0, i_2 = 1) \vee (i_1 = 1, i_2 = 0) \implies out$ = previous output state, as shown in Fig. 3.1. As a result, an observer monitoring the transition of output signal in a C-element can conclude that signal *out* changing from 0 to 1 $\implies i_1 = i_2 = 1$. Similarly, signal *out* changing from 1 to 0 $\implies i_1 = i_2 = 0$. Section 3.2.2 discusses the use of C-elements to detect completion of an event and indicate data validity for dual-rail circuits.

Fig. 3.1 Muller's C-element

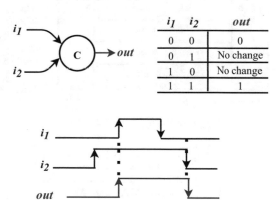

i_1	i_2	*out*
0	0	0
0	1	No change
1	0	No change
1	1	1

3.2.2 Dual-Rail Protocol

The dual-rail coding is the most popular M-of-N encoding utilised in various recent applications such as [14–18], to mention a few. As discussed previously, dual-rail protocol contains the information of the completion of an event by utilising two wires to encode a single bit of information. The req signal is embedded with both the data wires: $\{d.t, d.f\}$ for true (logic 1) and false (logic 0) values of data, respectively.

A two-phase dual-rail protocol utilises transition signaling, and a pair of data wires $\{d.t, d.f\}$ provides a unique code for data, as shown in Fig. 3.2. Therefore, logic 0 and logic 1 are represented by $\{0, 1\}$ and $\{1, 0\}$, respectively (refer Table 3.1). Since the output is captured at each signal transition, a two-phase dual-rail protocol does not need a spacer, unlike its four-phase counterpart, as discussed next. A new code is received on the n-bit channel if exactly one wire has made a transition in each of the n wire pairings.

A four-phase dual-rail protocol utilises level encoding, and it also uses the pair of data wires $\{d.t, d.f\}$, as shown in Fig. 3.3. Code $\{d.t, d.f\} = \{0, 0\}$ represents the "spacer" and no valid data is transmitted during that period. Code $\{d.t, d.f\} = \{0, 1\}$ represents logic 0 and $\{d.t, d.f\} = \{0, 1\}$ represents logic 1, and they are called "valid" data. Code $\{d.t, d.f\} = \{1, 1\}$ represents "invalid" data, and it is not used in a four-phase dual-rail protocol. A direct transition between two valid codes is not allowed, and a valid code should always go through a spacer to another valid code.

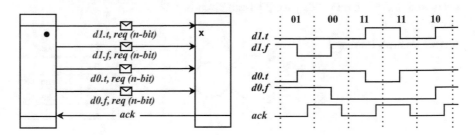

Fig. 3.2 Two-phase dual-rail protocol

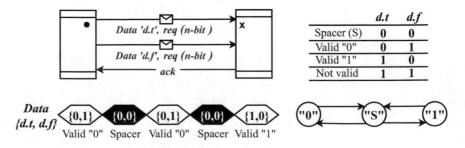

Fig. 3.3 Four-phase dual-rail protocol

Fig. 3.4 Four-Phase
Dual-Rail circuit with two
parallel paths

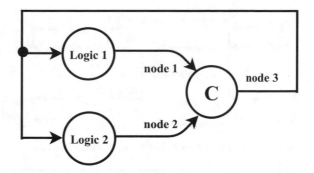

Muller's C-elements are used to determine the completion of an event and data validity in dual-rail circuits and are independent of wire delays. For instance, Fig. 3.4 illustrates a digital circuit comprised of two logic components, one C-element, and three nodes. The output signal at node 3 will change state only when the signals at node 1 and node 2 are in the same state, irrespective of the timing relation between node 1 and node 2. Therefore, C-element assists the circuit to assert the *ack* signal after the completion of an event, without getting affected by the circuit's initial state or the difference between the delay of two parallel paths.

Figure 3.5 represents an *n*-bit asynchronous dual-rail circuit using C-element for detecting valid data. The *req* signal is embedded with the input data *I*, and a single bit data I_i is represented using two wires $I_i.t$ (logic 1) and $I_i.f$ (logic 0), where $i < n$. The dual-rail circuit generates a spacer whenever the *req* signal is low, whereas a valid input data (refer to Fig. 3.3) sets the *req* signal high, and the circuit can start the computation. The final acknowledge signal *ack* indicates the completion of an event and should be asserted only after each bit of the output signal out_i has been validated by its corresponding acknowledge signal ack_i. Several OR gates are required to detect the valid output by ORing the $out_i.t$ and $out_i.f$ and generate the corresponding acknowledge signal ack_i. All these ack_i signals are given to an *n*-bit wide C-element that assists to generate the final *ack* signal and indicates the data validity.

The dual-rail circuits utilise two wires per bit which makes these circuits more complex than the synchronous design and bundled data approach. Using C-elements to detect valid data adds more complexity to the circuit, as it requires several OR logic gates to generate the individual acknowledge signal, and a wide C-element to collect these individual acknowledge signals and generate the final acknowledge signal *ack*. This additional complexity introduces additional delay to the circuit and detecting the data validity may need more delay than the actual delay of the digital computation block itself. Moreover, since dual-rail implementation needs

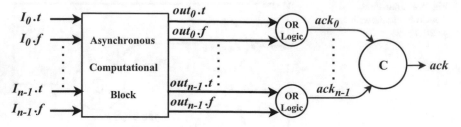

Fig. 3.5 n-bit asynchronous dual-rail circuit using C-element for completion detection

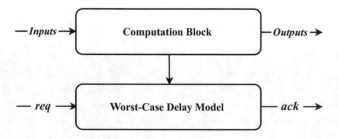

Fig. 3.6 Standard Bundled Data Channel with a worst-case delay

two wires per bit, and the completion detection scheme adds several OR gates and a wide C-element, the circuit consumes more area and power than its synchronous equivalent. The key disadvantage of this approach, however, is that the completion detection process starts after the output data is ready, introducing additional delay to the overall circuit.

3.3 Bundled Data Protocol

Bundled data protocol is defined as an arrangement where n-bit data is bundled with the req and ack signals. The term "*single-rail*" is also used for bundled data protocol, as it requires a single wire to represent one bit of data. A worst-case delay model is used to ensure data validity and the completion of an event in bundled data circuits, as shown in Fig. 3.6. The waveform for a bundled data channel using two-phase and four-phase signaling protocol is shown in Fig. 3.7a and b, respectively. In the two-phase signaling protocol, every transition represents an event, whereas the four-phase signaling protocol represents an event only for high logic level, and the handshaking signals must return to zero before commencing a new event. Thus, the four-phase bundled data protocol needs additional energy and time due to the return-to-zero transitions, whereas implementing the two-phase bundled data protocol is complex, and it is difficult to comment that which protocol is optimal.

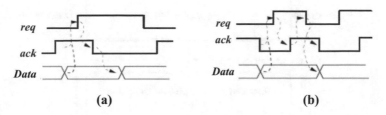

Fig. 3.7 Bundled data channel waveform using (**a**) two-phase signaling protocol and (**b**) four-phase signaling protocol

3.3.1   Completion Detection in Bundled Data Circuits

The bundled data protocol can be easily implemented by replacing the clock signal(s) with handshaking signals and a worst-case delay model. Bundled data protocol utilizes single-rail coding, which is similar to synchronous design style but without the clock-related complexity of synchronous design. Single-rail configurations are preferred because they are known to be area and power efficient. The primary disadvantage of the traditional bundled data protocol is that indicates the completion of a computation after the worst-case delay of the circuit. The speed of a bundled data circuit can be improved if there is a way to detect the process completion after actual computation delay. Completion detection schemes are frequently used in bundled data circuits to indicate the completion of an event as soon as the computation is complete and a valid output is available.

Nowick [19, 20] utilised the event-driven property of asynchronous circuits and proposed a **speculative** completion detection scheme for asynchronous adders using bundled data protocol. The computation delay of the adder is determined by its input operands when using this speculative completion detection scheme. A similar concept is used to design an asynchronous barrel shifter, in which the computation delay required to shift the input data depends upon the shift amount [21, 22]. The speculative approach by Nowick has nearly 200 citations and is still widely referred by recent researchers, such as [23–27], to mention a few. Although Nowick's technique is older, it is preferable to acknowledge the originator of a concept, which is why this book refers to Nowick's original work rather than to more recent works in which the mechanism is largely unchanged.

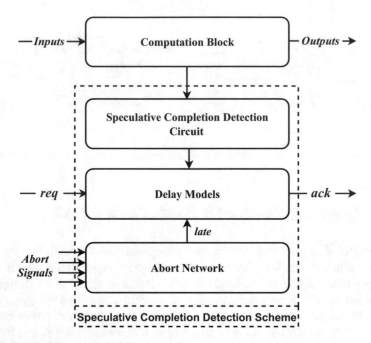

Fig. 3.8 Bundled data channel with a speculative completion detection scheme

A speculative completion detection scheme estimates the computation delay based on the input data and generates the *ack* signal after the estimated delay, as shown in Fig. 3.8. An abort network is used to abort the selected speculative path in case of incorrect delay estimation, and the *ack* signal is generated after the default worst-case delay. The speculative completion detection circuit operates parallel to the adder and generates the *ack* signal as soon as the output is available.

Figure 3.9 illustrates the speculative completion detection circuit utilised by Nowick [20] for asynchronous adders.

The completion detection circuit contains γ mutually exclusive signal paths with different delay models replicating the adder's delay for different input combinations, including the worst-case delay (delay model 1). Each speculative path, that is, the signal path with a delay smaller than the worst-case delay, is designed with an abort network. Once the input data arrives at the adder, the completion detector analyses the input data characteristics, estimates the computation delay, and suggests a delay path accordingly. The *ack* signal will be generated after the delay provided by the selected delay model. However, in case of late completion detection, a *late* signal must be asserted to abort the selected delay path and data is redirected to the worst-case delay.

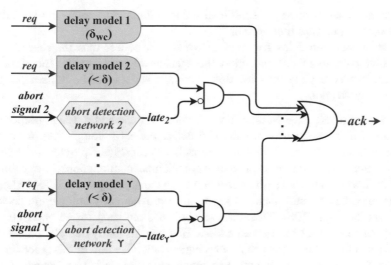

Fig. 3.9 Speculative Completion Detection Circuit [20]

> The term "late completion" refers to a condition in which the computation requires more computation delay than the selected speculative path. In contrast to the dual-rail protocol, which commences completion detection after the output becomes available, the speculative completion detection scheme operates parallel to the adder, and hence no additional delay is required to generate *ack* signal once the valid output is available.

The asynchronous adder designed using a speculative completion detection scheme has reported improved processing speed range from 5% to 30% over a comparable synchronous implementation. However, the performance of this approach is highly dependent upon the efficiency of abort network, as it should abort the selected speculative path before sending the acknowledge signal in the case of an incorrect delay model is selected. Lai [28] has found that there are few cases that are not detectable by the abort network under certain conditions and acknowledge signal has been sent before the computation is complete, resulting in an incorrect outcome. The speculative completion detection scheme has four major drawbacks:

1. It is unable to provide the correct information of completion for all input combinations, i.e., speculative computation can potentially fail to abort an early completion under certain conditions [28].
2. The selected speculative path corresponding to the ongoing event must be aborted if the speculation is incorrect, and the worst-case delay path is selected by default.

3. The abort network needs additional computation delay, more implementation area, and consumes more power.
4. The computation delay for abort network must be less than the selected speculative path so that it can abort the speculative path timely. Therefore, this technique is feasible only for slower circuits and cannot be utilised for high-speed applications.

Few researchers have worked towards improving the adder design proposed by Nowick. An approach was proposed to detect the early completion of an event for asynchronous adders [29]. This approach was capable of providing the correct information of the completion of an event for all input combinations, eliminating the need to abort the selected path corresponding to the ongoing event. This design was further modified by introducing a kill term to provide early completion detection for more input data [28]. Therefore, it eliminates the need for an abort network, and it can be used for high-speed adders. However, the early completion detection approach is limited to the asynchronous adders only. The aim of this research work is to propose a generic architecture to detect early completion and hence provides the deterministic completion detection scheme for any asynchronous computation block using bundled data protocol.

3.4 Analysis of Completion Detection Schemes

Completion detection circuits are commonly used in asynchronous circuits to indicate the completion of an event as soon as the computation is done, and a valid output is available. For instance, dual-rail protocol utilizes multiple OR gates and C-elements to detect the valid data, and the completion detection process in dual-rail protocol starts after the computation is done. Similarly, a typical bundled data circuit generates the ack signal after the worst-case delay δ_{wc}, but using a completion detection circuit can reduce this worst-case delay to average-case delay by generating the ack signal as soon as the valid output is available. A comparison of existing completion detection schemes is given in Fig. 3.10.

A. L. Davis developed a Data-Driven Machine (DDM) by using a completion detection approach [30], in which the synchronous circuit is converted into asynchronous by controlling the internal clock of the circuit. The internal clock is activated after receiving the request signal, and once the computation is done, an acknowledge signal is sent and the internal clock is deactivated. Although this approach works fine with large modules, it is not feasible for smaller modules. DDM utilise Data-Driven Network (DDN) [31], which is considered non-deterministic as it cannot determine any unique execution history. The logic required to control the internal clock is complex, and it fails to address the delay associated with starting the clock generator. Another approach known as Q-modules [32] is designed for delay-insensitive specifications that utilise internal clocks yet avoid the synchronisation failures. However, Q-modules require the clock phase to be greater than the longest

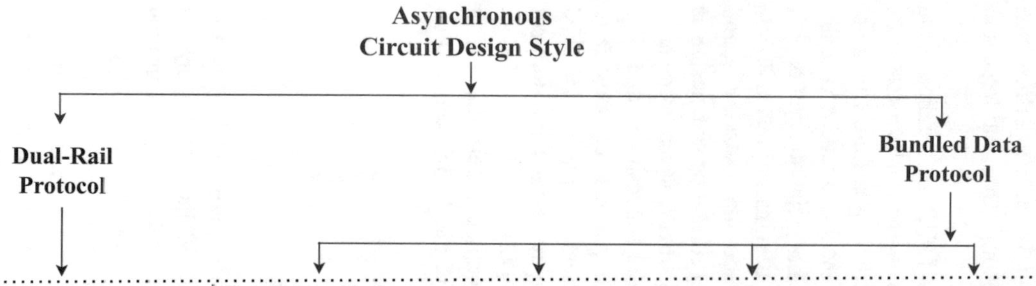

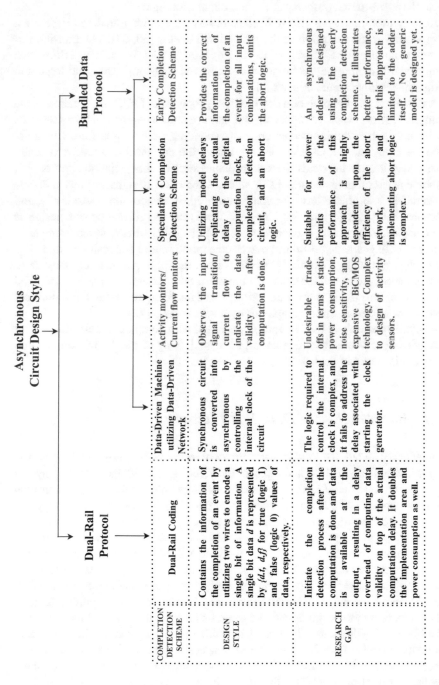

COMPLETION DETECTION SCHEME	Dual-Rail Coding	Data-Driven Machine utilizing Data-Driven Network	Activity monitors/ Current flow monitors	Speculative Completion Detection Scheme	Early Completion Detection Scheme
DESIGN STYLE	Contains the information of the completion of an event by utilizing two wires to encode a single bit of information. A single bit data *d* is represented by *{d.t, d.f}* for true (logic 1) and false (logic 0) values of data, respectively.	Synchronous circuit is converted into asynchronous by controlling the internal clock of the circuit	Observe the input signal transition/ current flow to indicate the data validity after computation is done.	Utilizing model delays replicating the actual delay of the digital computation block, a completion detection circuit, and an abort logic.	Provides the correct information of the completion of an event for all input combinations, omits the abort logic.
RESEARCH GAP	Initiate the completion detection process after the computation is done and data is available at the output, resulting in a delay overhead of computing data validity on top of the actual computation delay. It doubles the implementation area and power consumption as well.	The logic required to control the internal clock is complex, and it fails to address the delay associated with starting the clock generator.	Undesirable trade-offs in terms of static power consumption, noise sensitivity, and expensive BiCMOS technology. Complex to design of activity sensors.	Suitable for slower circuits as the performance of this approach is highly dependent upon the efficiency of the abort network, and implementing abort logic is complex.	An asynchronous adder is designed using the early completion detection scheme. It illustrates better performance, but this approach is limited to the adder itself. No generic model is designed yet.

Fig. 3.10 Analysis of completion detection schemes

delay or the worst-case delay of the digital computation block and hence cannot utilise the event-driven property of asynchronous paradigm.

Activity monitors were also utilised to observe the input signal and generate a pulse whenever the transition of the input signal occurs [33, 34]. The transition of the input signal can be either low to high or high to low. Later, the performance of the activity monitoring technique was improved by using a dynamic logic and unidirectional input signal transition [35].

A novel design of completion detector was proposed in [36], which indicates the completion of an event by observing the current flow. The current flow increases significantly when the input data and corresponding output values are changing. Once the input and output variables are stable, the current flow reduces virtually to zero, and the circuit sends the acknowledge signal. This technique works well for some asynchronous circuits [37], whereas for other designs, it presents some undesirable trade-offs in terms of static power consumption, noise sensitivity, and expensive BiCMOS technology [38, 39]. Moreover, the current sensor design is also complex. A similar approach was proposed [36, 40] where the requirement of reference voltages leads to higher power consumption and expensive design. The approach proposed by Gamble [41] suffers from the high quiescent current due to the reference current requirement of currents mode current comparator. Bundled data protocol was used for error detection and adaptive delay model, but it does not discuss how to ensure that a computation process is done [42].

Another approach to implementing a bundled data completion detection circuit is by utilising model delays replicating the actual delay of the digital computation block, as discussed in Sect. 3.3.1.

3.4.1 Bundled Data vs Dual-Rail Protocol

As previously stated, the bundled data protocol employs single-rail coding similar to synchronous circuits [43] and replaces the clock signal with handshaking signals and the worst-case delay model. Therefore, although bundled data circuits consume the same amount of silicon area and power as synchronous systems, they are unable to take advantage of the event-driven property of the asynchronous paradigm [44, 45]. On the contrary, the dual-rail design encodes the information of data validity within the data and indicates the data validity after actual computation delay. The receiver can identify the completion of an ongoing event as soon as it detects the valid data, and ideally, the dual-rail circuits are faster than bundled data or synchronous circuits. However, the process of determining data validity in dual-rail circuits requires a large number of gates, as each bit in the signal path needs to be examined. The logic to determine data validity requires either a wide C-element or multiple stages of smaller C-elements that needs a considerable computational delay, and hence, it does not deliver the anticipated outcome [46, 47]. The main drawback of dual-rail circuits is that they initiate the completion detection process

after the computation is complete and the data is available at the output, resulting in a delay overhead of computing data validity on top of the actual computation delay. Moreover, with dual-rail implementation requiring a double implementation area, it creates a challenge for applications where silicon area is a major concern, and the massive data validity logic required in the dual-rail protocol leads to higher power consumption as well.

The completion detection schemes in bundled data protocol operates parallel to the ongoing computation, which avoids delay overhead of computing data validity on top of the actual computation delay. However, existing completion detection schemes for bundled data protocol have several drawbacks, as explained in Sect. 3.3.1 and illustrated in Fig. 3.10. The asynchronous adder designed using early completion detection scheme with bundled data protocol demonstrated improved response compared to their synchronous counterpart, and hence a generic model to utilise a deterministic completion detection scheme is proposed to develop in this research.

3.5   Chapter Summary

This chapter provides a review of literature on existing completion detection schemes for asynchronous paradigm. The M-of-N protocols and its few special cases are explored in this chapter. A brief introduction of Muller's C-element is provided, and its application to detecting the completion of an event in dual-rail circuits is described. A standard bundled data circuit with a worst-case delay model is explained, and its performance is compared with dual-rail circuits. The chapter highlights the importance of having an efficient completion detection scheme and reviews different existing completion detection schemes. The chapter analyses a speculative completion detection scheme that employs various delay models and an abort network. Additionally, this chapter addresses the drawbacks of the speculative completion detection scheme and analyses a few existing works that implement the asynchronous adder with an early completion detection scheme without using an abort network.

References

1. Toosizadeh, N. (2010), Enhanced synchronous design using asynchronous techniques, PhD thesis, University of Toronto.
2. Naqvi, Syed Rameez, Robert Najvirt, and Andreas Steininger (2013). "A multi-credit flow control scheme for asynchronous NoCs". In: 2013 IEEE 16th International Symposium on Design and Diagnostics of Electronic Circuits & Systems (DDECS). IEEE, pp. 153–158.
3. Clark, Wesley A (1967). "Macromodular computer systems". In: Proceedings of the April 18-20, 1967, spring joint computer conference. ACM, pp. 335–336.

4. Stucki, Mishell J, SeveroMOrnstein, and WesleyAClark (1967). "Logical design of macro-modules". In: Proceedings of the April 18-20, 1967, spring joint computer conference. ACM, pp. 357–364.
5. Verhoeff, Tom (1994). "A theory of delay-insensitive systems". In: Technische Universiteit Eindhoven.https://doi.org/10.6100/IR416309
6. Dean, M. E. (1991), Efficient self-timing with level-encoded 2-phase dual-rail (LEDR), in 'Proceedings of University of California/Santa Cruz conference on Advanced research in VLSI, 1991', pp. 55–70.
7. McGee, P. B., Agyekum, M. Y., Mohamed, M. A. and Nowick, S. M. (2008), A level encoded transition signaling protocol for high-throughput asynchronous global communication, in '2008 14th IEEE International Symposium on Asynchronous Circuits and Systems', IEEE, pp. 116–127.
8. Tabassam, Z., Naqvi, S. R., Akram, T., Alhussein, M., Aurangzeb, K. and Haider, S. A. (2019), 'Towards designing asynchronous microprocessors: From specification to tapeout', IEEE Access 7, 33978–34003.
9. Sobelman, G. E. and Fant, K. (1998), CMOS circuit design of threshold gates with hysteresis, in 'ISCAS'98. Proceedings of the 1998 IEEE International Symposium on Circuits and Systems (Cat. No. 98CH36187)', Vol. 2, IEEE, pp. 61–64.
10. Fant, K. M. and Brandt, S. A. (1997), 'Null convention logic', Technical document, http://www.theseus.com/PDF/nclpaper.pdf.
11. Fant, K. and Brandt, S. (1996), 'Null convention logic: a complete and consistent logic for asynchronous digital circuit synthesis proceedings of international conference on application specific systems', Architectures and Processors.
12. Sparsø, J. (2020), Introduction to Asynchronous Circuit Design., DTU Compute, Technical University of Denmark.
13. Muller, David E (1956). "A theory of asynchronous circuits". In: Report 75, University of Illinois.
14. Wheeldon, A., Morris, J., Sokolov, D. and Yakovlev, A. (2019), 'Self-timed, minimum latency circuits for the internet of things', Integration 69, 138–146.
15. Lin, Y. and Cheng, C. (2019), 'Dual rail device with power detector'. US Patent App. 16/181,889.
16. Su, J. (2019), 'Full-swing dual-rail SRAM sense amplifier'. US Patent App. 15/839,375. Sutherland, I. E. (1989), 'Micropipelines', Communications of the ACM 32(6), 720–738.
17. Shi, Y., Schulze, T., Kwiat, K. and Kamhoua, C. (2019), 'Security method for resisting hardware trojan induced leakage in combinational logics'. US Patent App. 16/174,442.
18. Jain, S. K. and Katoch, A. (2019), 'Dual rail SRAM device'. US Patent App. 16/176,168.
19. Nowick, S. M., Yun, K. Y., Beerel, P. A. and Dooply, A. E. (1997), Speculative completion for the design of high-performance asynchronous dynamic adders, in 'Proceedings Third International Symposium on Advanced Research in Asynchronous Circuits and Systems', IEEE, pp. 210–223.
20. Nowick, S. M. (1996), 'Design of a low-latency asynchronous adder using speculative completion', IEE Proceedings-Computers and Digital Techniques 143(5), 301–307.
21. Beerel, P. A., Kim, S., Yeh, P.-C. and Kim, K. (1999), Statistically optimized asynchronous barrel shifters for variable length codecs, in 'Proceedings. 1999 International Symposium on Low Power Electronics and Design (Cat. No. 99TH8477)', IEEE, pp. 261–263.
22. Beerel, P. A. and Kim, K.-S. (2003), 'Statistically optimized asynchronous barrel shifters for variable length codecs', The Journal of Korean Institute of Communications and Information Sciences 28(11A), 891–901.
23. Thakur, G., Sohal, H. and Jain, S. (2021), 'A novel ASIC-based variable latency speculative parallel prefix adder for image processing application', Circuits, Systems, and Signal Processing 40, 5682–5704.
24. Esposito, D., De Caro, D. and Strollo, A. G. M. (2016), 'Variable latency speculative parallel prefix adders for unsigned and signed operands', IEEE Transactions on Circuits and Systems I: Regular Papers 63(8), 1200–1209.

25. Cilardo, A. (2015), Variable-latency signed addition on FPGAs, in '2015 25th International Conference on Field Programmable Logic and Applications (FPL)', IEEE, pp. 1–6.
26. Esposito, D., De Caro, D., Napoli, E., Petra, N. and Strollo, A. G. M. (2015), 'Variable latency speculative Han-Carlson adder', IEEE Transactions on Circuits and Systems I: Regular Papers 62(5), 1353–1361.
27. Hand, D., Cheng, B., Breuer, M. and Beerel, P. A. (2015), 'Blade–a timing violation resilient asynchronous design template'.
28. Lai, K. K. (2016), Novel Asynchronous Completion Detection For Arithmetic Datapaths, Taylor's University, Malaysia.
29. Koes, D., Chelcea, T., Onyeama, C. and Goldstein, S. C. (2005), Adding faster with application specific early termination, Technical report, Carnegie-Mellon Univ Pittsburgh PA publisher of Computer Science.
30. Davis, A. L. (1978), The architecture and system method of ddm1: A recursively structured data driven machine, in 'Proceedings of the 5th annual symposium on Computer architecture', ACM, pp. 210–215.
31. Davis, A. (1974), 'Data driven nets–a class of maximally parallel, output-functional program schemata', Burroughs IRC Report, San Diego.
32. Rosenberger, F. U., Molnar, C. E., Chaney, T. J. and Fang, T.-P. (1988), 'Q-modules: Internally clocked delay-insensitive modules', IEEE transactions on computers 37(9), 1005–1018.
33. Grass, E. and Jones, S. (1996), 'Activity-monitoring completion-detection (AMCD): A new approach to achieve self-timing', Electronics Letters 32(2), 86–88.
34. Grass, E., Morling, R. C. and Kale, I. (1996), Activity-monitoring completion-detection (AMCD): A new single rail approach to achieve self-timing, in 'Proceedings Second International Symposium on Advanced Research in Asynchronous Circuits and Systems', IEEE, pp. 143–149.
35. Bartlett, V. and Grass, E. (1997), 'Completion-detection technique for dynamic logic', Electronics Letters 33(22), 1850–1852.
36. Dean, M. E., Dill, D. L. and Horowitz, M, (1994), 'Self-timed logic using current-sensing completion detection (CSCD)', Journal of VLSI signal processing systems for signal, image and video technology 7(1-2), 7–16.
37. Akgun, O. C., Leblebici, Y. and E. A., V. (2007), Design of completion detection circuits for self-timed systems operating in subthreshold regime, in '2007 Ph. D Research in Microelectronics and Electronics Conference', IEEE, pp. 241–244.
38. Izosimov, O., Shagurin, I. and Tsylyov, V. (1990), 'Physical approach to CMOS module self-timing', Electronics Letters 26(22), 835–836.
39. Grass, E. and Jones, S. (1995), Asynchronous circuits based on multiple localised current sensing completion detection, in 'Proceedings Second Working Conference on Asynchronous Design Methodologies', IEEE, pp. 170–177.
40. Varshavsky, V. I., Marakhovsky, V. and Lashevsky, R. (1995), Asynchronous interaction in massively parallel computing systems, in 'Proceedings 1st International Conference on Algorithms and Architectures for Parallel Processing', Vol. 2, IEEE, pp. 481–492.
41. Gamble, M. J. (1995), A novel current-sensing completion-detection circuit adapted to the micropipeline methodology, University of Manitoba, Canada.
42. Gimenez, G., Simatic, J. and Fesquet, L. (2020), An adaptive delay error asynchronous circuit design based on bundled-data handshake protocol, in '2020 7th International Forum on Electrical Engineering and Automation (IFEEA)', IEEE, pp. 596–600.
43. Rahbaran, B. and Steininger, A. (2008), 'Is asynchronous logic more robust than synchronous logic?', IEEE Transactions on dependable and secure computing 6(4), 282–294.
44. Srivastava, P., Chung, E. and Ozana, S. (2020), 'Asynchronous floating-point adders and communication protocols: A survey', Electronics 9(10), 1687.
45. Srivastava, P. and Chung, E. (2022), 'An asynchronous bundled-data barrel shifter design that incorporates a deterministic completion detection technique', IEEE Transactions on Circuits and Systems II: Express Briefs 69(3), 1667–1671.

46. Jing, Z., Guochen, D., Binbin, S. and Fang, Z. (2020), An adaptive delay error asynchronous circuit design based on bundled-data handshake protocol, in '2020 7th International Forum on Electrical Engineering and Automation (IFEEA)', IEEE, pp. 596–600.
47. Poole, N. (1994), 'Self-timed logic circuits', Electronics & communication engineering journal 6(6), 261–270.

Chapter 4
Case Studies: Barrel Shifter and Binary Adders

4.1 Barrel Shifter

Shifting is a fundamental logical operation that is used in a variety of computing machines and processors, including Arithmetic Logic Units (ALUs), Floating-Point Units (FPUs), Graphics Processing Units (GPUs), Digital Signal Processing (DSP) units, etc. The simplest way to implement the shifter circuit is by using flip-flops, shifting one bit at a time, but it is not acceptable for massive shifts due to the need for a long computation delay. One strategy for implementing a shifter with a massive shift amount is to design it in such a way that it can shift a certain number of bits in a single clock cycle [1]. The documentation for the CDC (Control Data Corporation) 6600 computer from the early 1970s has one of the first mentions of such a shifter architecture [2]. A similar design of shifter circuit was utilised for the first time [3] in the Numeric Data Processor, known as Barrel Shifter, that can provide multi-bit shifting in a single operation [4]. The term *"barrel"* in the barrel shifter is derived from the fact that when a gun's barrel is rotated, its entire contents shift in sync. Barrel shifters can be implemented by using a multiplier or by using a logarithmic shifter. The multiplier-based barrel shifters are considered only if the multiplier is embedded in the device. The logarithmic shifter is a preferred way to implement the barrel shifter for a large shifting amount, as it is almost two times faster than multiplier-based shifters [5, 6]. The next section describes the fundamental design of a conventional barrel shifter (CBS). Understanding the fundamental architecture of CBS will aid in comprehension of the architecture developed for the barrel shifter in the following chapter. The notation used in this book to understand the concept of barrel shifter is listed in the following paragraph.

Notation
Variables $S, i, n, s, j, k, \lambda \in \mathbb{Z}_+$; S_i denotes the select line for all multiplexers present in stage i of an n-bit CBS. Variable s represents the shift amount of an n-bit CBS such that $0 \leq s < n$. The variable λ denotes the total number of shift stages

P. Srivastava, *Completion Detection in Asynchronous Circuits*,
https://doi.org/10.1007/978-3-031-18397-3_4

required to shift the input data I by s-bits using an n-bit CBS. $I = I_0 I_1 \ldots I_{n-1}$ represents an n-bit of input data to be shifted, and $Y = Y_0 Y_1 \ldots Y_{n-1}$ represents the corresponding shifted n-bit output data.

4.1.1   Architecture of a Conventional Barrel Shifter (CBS)

The architecture of an n-bit logarithmic CBS implemented using 2×1 multiplexers is discussed in this section. A single CBS stage i is implemented using n 2×1 multiplexers, and hence the total number of 2×1 multiplexers required to implement an n-bit CBS is $n \times \lambda$, where λ is the total number of CBS stages and $\lambda = log_2 n$. The input data I is always connected to the CBS stage 0, and the final shifted output is captured from the CBS stage $n - 1$. The input of an intermediate stage i is connected to the output of its previous stage $i - 1$, and its output is connected to the input of its next stage $i + 1$. The logic diagram of an 8-bit CBS to perform a right shift utilising 2×1 multiplexers is shown in Fig. 4.1, and any n-bit right/left shift CBS can be designed using the similar concept. The 2×1 multiplexers used to implement the CBS are denoted by $m_{j:k}$, which represents the jth multiplexer of the stage k, and $0 \leq j \leq n - 1$.

Figure 4.2 illustrates a critical path of an 8-bit right-shift CBS; it is a more abstract representation of the 8-bit CBS design, and it is easier to illustrate an n-bit

Fig. 4.1 Logic Diagram of an 8-bit Right-Shift Conventional Barrel Shifter (CBS)

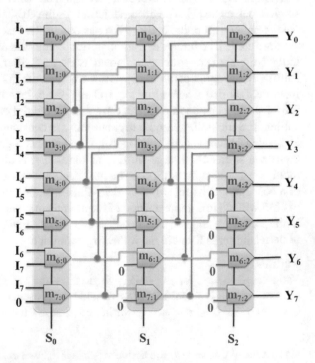

CBS in this fashion when n is large. An individual CBS stage i can either shift the input data I by a certain amount or pass the data to the next stage without shifting, depending upon the value of the select line S_i corresponding to a shift amount s.

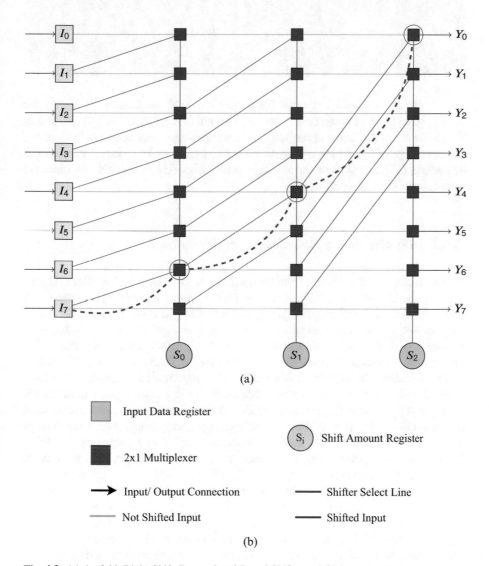

(a)

Input Data Register

S_i Shift Amount Register

2x1 Multiplexer

→ Input/ Output Connection

—— Shifter Select Line

—— Not Shifted Input

—— Shifted Input

(b)

Fig. 4.2 (a) An 8-bit Right-Shift Conventional Barrel Shifter and (b) its components

The shift amount s can be expressed as follows [7] in terms of the select lines of the CBS stages:

$$s = \sum_{i=0}^{\lambda-1} S_i.2^i.$$ (4.1)

It is clear from Eq. 4.1 that a CBS stage i can shift the input I by 2^i bits for $S_i = 1$, else it will pass the data to its output without shifting for $S_i = 0$. All the λ-stages are cascaded such that the output of a shifter stage is connected to the input of the next stage of the shifter, and multi-bit shifting in a single operation is achieved. Therefore, CBS can perform the shifting operation quickly and provide the outcome with less programming effort [8].

4.1.2 Asynchronous Bundled Data Barrel Shifter

The architecture of a 16-bit Asynchronous Bundled Data Barrel Shifter (ABBS) using a speculative completion detection method was proposed in [9], similar to Nowick's idea of speculative completion detection for the asynchronous adder [10]. The proposed architecture utilises two-level barrel shifters BS1 and BS2. It is also known as 6-10 design, i.e., BS1 alone is used for six most common (smaller) shifts and ten other uncommon (larger) shifts must pass through BS1 and BS2. Yeh et al. [11] explained the algorithm to identify the common and uncommon shifts for a 16-bit barrel shifter. The 6-10 design reduces the energy consumption significantly for a given average performance, compared to their synchronous counterpart, but it needs an abort network to prevent false triggering of early completion. Even though using an abort network is a sufficient condition, it is not a necessary condition for bundled data circuits [8], and hence the need for an abort network can be eliminated with a superior completion detection scheme. A similar approach is proposed recently for a two-stage synchronous barrel shifter, and the shift amount is divided into two steps to reduce the timing delay [12]. Several recent articles continue to contribute to the performance improvement of the shifter [13–18] and its applications [19–24]. Hence, a barrel shifter was selected to validate the proposed architecture, and an alternate design of ABBS was developed using a deterministic completion detection scheme.

4.2 Binary Adders

The adder is a fundamental unit of any arithmetic unit, as addition and subtraction are the most frequently performed arithmetic operations in most scientific applications [25]. Adders are also used to implement more complex operations such as multiplication, division, and other transcendental operations. The Ripple Carry Adder (RCA) is the simplest implementation of a binary adder. It is built by cascading multiple full adders (FA) and needs long computation time to calculate the final output. The long computation time in RCA due to carry generation in each phase is reduced in the Carry Look Ahead Adder (CLA) by generating carry ahead of time. Appendix 4.4 details the architecture of RCA and CLA. A larger bit CLA suffers from a larger delay because the carry generation through the look-ahead block consumes more time. It can be reduced by overlapping the look-ahead blocks. A multilevel tree of look-ahead structures can be created in order to obtain a delay that increases with $log_2 n$, where n is the adder width [26–28]. These adders are known as Parallel Prefix Adder (PPA). The architecture of PPA can be categorised into three stages, as explained here and graphically in Fig. 4.3:

1. **Pre-processing Stage:** Generation of Group Generate and Group Propagate Signal.
2. **Carry generation Stage:** Generation of Carry Signal using Group Generate and Group Propagate Signal.
3. **Post-processing Stage:** Generation of Sum bit using Group Generate, Group Propagate, and Carry Signal.

PPAs are typically represented using their fundamental blocks, i.e., black cells, grey cells, and white cells to depict the propagate, generate, and buffer logics

Fig. 4.3 Parallel Prefix Adder Stages

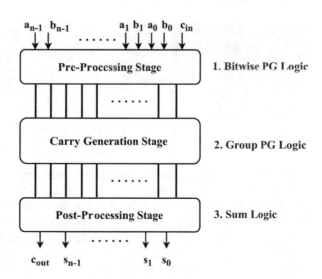

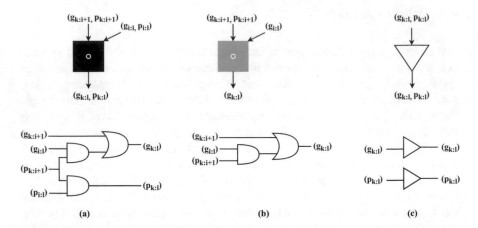

Fig. 4.4 Fundamental blocks of PPA: (**a**) Black Cell, (**b**) Grey Cell, and (**c**) White Cell (Buffer)

[29, 30]. Figure 4.4 illustrates the logic of these cells, followed by their logical expressions in Eq. 4.2.

$$g_{k:l} = g_{k:i+1} + p_{k:i+1} \cdot g_{i:l},$$

$$p_{k:l} = p_{k:i+1} \cdot p_{i:l}, \tag{4.2}$$

where $k > i > l$, and $g_{k:l}$ and $p_{k:l}$ are the group generate and propagate signals, respectively.

The addition of ith bit is capable of generating a carry if both the input operands are 1, providing the carry generate signal $g_i = a_i \cdot b_i$. Similarly, the addition of ith bit is capable of propagating a carry if either one of the input operands is 1, providing the carry propagate signal $p_i = a_i \oplus b_i$. The kill term is defined such that $k_i^0 = (a_i + b_i)'$. Kindly refer Appendix 4.4 for a detailed explanation of generate and propagate signals.

There are various designs proposed by researchers to implement PPA, based on the number of logic gates, the number of fan-out from each gate, the number of wiring between levels, and the number of levels of the logic [31, 32]. Most commonly used parallel prefix adder architectures are Brent-Kung, Sklansky, Kogge-Stone, Han-Carlson, and Ladner-Fischer. Brent-Kung Adder [30] has the maximum logic depth, and it is the slowest adder among the others. Kogge-Stone [35] and Sklansky Adders [36] have the lowest logic depth. The performance of

Table 4.1 Comparing different architectures of a 32-bit PPA [33]

| | Computation nodes | | |
Adder type	Black cell	Grey cell	Logic depth
Brent-Kung	26	31	8
Kogge-Stone	98	31	5
Han-Carlson	33	31	6
Ladner-Fischer	33	31	6
Skalansky	33	31	5

Han-Carlson [37] and Ladner-Fischer [38] adders lies in between. A comparison between different adder implementation is provided in Table 4.1 [33, 34].

4.2.1   Asynchronous Bundled Data Adder

The choice of PPA architecture for an application depends upon the completion detection technique used to design Asynchronous PPA. Nowick et al. [10, 39] have proposed a speculative completion detection technique to implement the Asynchronous Brent-Kung (BK) adder using bundled data protocol. Please refer to Appendix 4.4 for the Brent-Kung adder architecture. Although the technique proposed by Nowick is an old publication, it has nearly 200 citations and is still widely referred by recent researchers, such as [40–44], and the core principle of speculative completion detection remains largely unchanged. Therefore, Nowick's original work is referred in this book for speculative completion detection technique for adders.

The speculative completion detection scheme for asynchronous PPA is implemented with multiple delay models: one worst-case delay model and one or more speculative matching delay models to indicate early completion (refer to Fig. 3.9). The selection of a delay model is operand dependent, and a matched delay model is selected based upon the input operand characteristics for an addition operation. The speculative method utilises the *late* detection logic, which identifies the process demanding more processing time than the selected speculative delay model to provide the final output. An abort detection network is used to implement the *late* detection logic by terminating the false triggering for early completion detection, and the data is redirected to the worst-case delay. Nowick [39] presented a necessary condition that a run of eight consecutive Level-0 propagate signals must exist for late completion detection.

For a 32-bit BK adder, the condition for late completion detection in speculative completion detection scheme is defined as $p_0 p_1 p_2 p_3 p_4 p_5 p_6 p_7 + p_1 p_2 p_3 p_4 p_5 p_6 p_7 p_8 + \cdots + p_{24} p_{25} p_{26} p_{27} p_{28} p_{29} p_{30} p_{31}$.

The logic depth of an abort network is calculated as $(log_2 K + log_2 N)$ [29], where K denotes the total number of terms present in the *late* signal, and N is the total number of p signals in each term. Therefore, for eight consecutive Level-0 propagate signals, with K = 25 and N = 8, logic depth is 8; hence, the delay of abort network is comparable to that of the Brent-Kung adder, the slowest PPA in Table 4.1. The abort network must be faster than the adder itself, otherwise the adder will always operate at the worst-case delay. Therefore, the condition is safely approximated with smaller values of K and N, to produce a simplified logic. The approximations are done with 3-literal products, 4-literal products, and 5-literal products.

The 3-literal product equation is given as $p_5 p_6 p_7 + p_{11} p_{12} p_{13} + p_{17} p_{18} p_{19} + p_{23} p_{24} p_{25} + p_{29} p_{30} p_{31}$, and it consists of 5 product terms. The 4-literal/5-product is given by the equation $p_4 p_5 p_6 p_7 + p_9 p_{10} p_{11} p_{12} + p_{14} p_{15} p_{16} p_{17} + p_{19} p_{20} p_{21} p_{22} + p_{24} p_{25} p_{26} p_{27}$, and 5-literal/7-product equation is given by $p_3 p_4 p_5 p_6 p_7 + p_7 p_8 p_9 p_{10} p_{11} + p_{11} p_{12} p_{13} p_{14} p_{15} + p_{15} p_{16} p_{17} p_{18} p_{19} + p_{19} p_{20} p_{21} p_{22} p_{23} + p_{23} p_{24} p_{25} p_{26} p_{27} + p_{27} p_{28} p_{29} p_{30} p_{31}$.

A 4-literal/5-product network is utilised to implement a BK adder that gives the logic depth 5; oversimplifying the network does not provide accurate results, rendering this method incompatible for fast adder architectures with lower logic depth. Skalansky and Kogge-Stone adders have the lowest logic depth, as given in Table 4.1. However, since the Skalansky adder requires fewer cells than the Kogge-Stone adder, a Sklansky adder architecture is utilised to validate the proposed generic completion detection scheme.

4.3 Chapter Summary

This chapter covers the fundamentals of addition and shift operations, as these are the two main functions of ALU. The chapter discusses the architecture of barrel shifter and binary adder, as the proposed generic architecture of deterministic completion detection scheme is validated using asynchronous bundled data barrel shifter and binary adder.

Appendix

4.4 Fundamentals of Binary Adders

The adder is a fundamental unit of any arithmetic unit, as addition and subtraction are the most frequently performed arithmetic operations in most scientific applications. The fundamental binary adder is a Ripple Carry Adder, as discussed in the next section.

4.4.1 Ripple Carry Adder (RCA)

The Ripple Carry Adder (RCA) is the simplest implementation of a binary adder. It is built by cascading multiple full adders (FA), as shown in Fig. 4.5. a_k and b_k are the two input operands, c_k is the carry signal at any stage $k > 0$, and c_0 is the initial value of the carry signal. The RCA needs long computation time to calculate the final output, as the carry signal from each stage ripples through the adder from the Least Significant Bit (LSB) to the Most Significant Bit (MSB).

4.4.2 Carry Look Ahead Adder (CLA)

The long computation time in RCA due to carry generation in each phase is reduced in the Carry Look Ahead Adder (CLA) by generating carry ahead of time. The CLA creates two signals: Carry Generate Signal (g) and Carry Propagate Signal (p). The

Fig. 4.5 Ripple Carry Adder

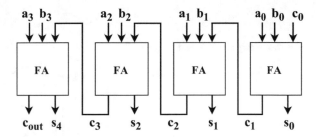

g and p signals are defined such that it can convey the information of the carry signal c for each bit position, as given by the following equations [29]:

$$g = a \cdot b,$$
$$p = a \oplus b,$$
$$c = a \cdot b + (a \oplus b) \cdot c \qquad (4.3)$$
$$= g + p \cdot c,$$
$$sum = a \oplus b \oplus c.$$

The logical circuit for the carry generate and carry propagate signals of a single bit adder is illustrated in Fig. 4.6, and any n-bit CLA can be designed using this fundamental circuit. The generate, propagate, and carry signals are evaluated concurrently in the CLA architecture, which enables the CLA to leverage this concurrency to unroll the carry recurrence. For instance, equations of carry signals for a 4-bit CLA can be given by the following equation:

$$c_1 = g_0 + p_0 \cdot c_0,$$
$$c_2 = g_1 + p_1 \cdot c_1$$
$$= g_1 + p_1 \cdot g_0 + p_1 \cdot p_0 \cdot c_0,$$
$$c_3 = g_2 + p_2 \cdot c_2$$
$$= g_2 + p_2 \cdot g_1 + p_2 \cdot p_1 \cdot g_0 + p_2 \cdot p_1 \cdot p_0 \cdot c_0,$$
$$c_4 = g_3 + p_3 \cdot c_3$$
$$= g_3 + p_3 \cdot g_2 + p_3 \cdot p_2 \cdot g_1 + p_3 \cdot p_2 \cdot g_1 \cdot g_0 + p_3 \cdot p_2 \cdot p_1 \cdot p_0 \cdot c_0.$$
$$(4.4)$$

Fig. 4.6 Carry Look Ahead Adder

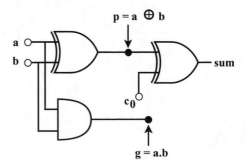

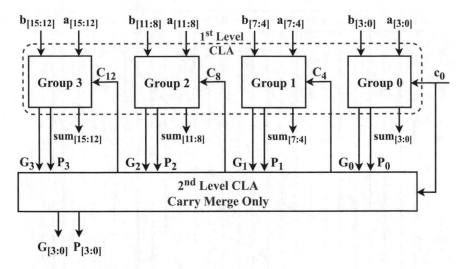

Fig. 4.7 A two-level 16-bit CLA [29]

A CLA with a large data width can be implemented by cascading CLA modules with a smaller data width. Figure 4.7 shows how four 4-bit CLAs can be used to create a 16-bit wide CLA. The generate and propagate signals of each group are $G_0, P_0, G_1, P_1, G_2, P_2, G_3$, and P_3, respectively. The group generate and propagate signals for a 16-bit CLA are denoted by $G_{[3:0]}$ and $P_{[3:0]}$, respectively. The equations for carry signals C_4, C_8, and C_12, produced from second-level CLA, are as follows [29]:

$$C_4 = G_0 + c_0 \cdot P_0,$$
$$C_8 = G_1 + G_0 \cdot P_1 + c_0 \cdot P_0 \cdot P_1, \qquad (4.5)$$
$$C_{12} = G_2 + G_1 \cdot P_2 + G_0 \cdot P_1 \cdot P_2 + c_0 \cdot P_0 \cdot P_1 \cdot P_2.$$

The two-level 16-bit CLA performs the following operations:

1. All the 4-bit CLAs at the first level generate the individual (bitwise) generate (g_i) and propagate (p_i) signals. Group generate ($G_{i:j}$) and Group propagate ($P_{i:j}$) signals are produced concurrently.
2. The second level of CLA generates carry signals C_4, C_8, and C_{12}, which are fed back into the 4-bit CLAs at the first level into Group 1, Group 2, and Group 3, respectively, and the individual sum of 4-bit CLAs is evaluated.

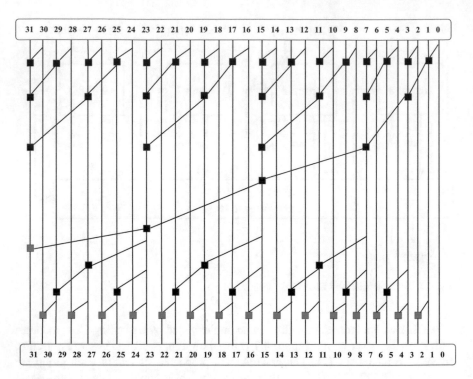

Fig. 4.8 Brent-Kung Adder

4.4.3   *Brent-Kung Adder*

Parallel Prefix Adders (PPAs) used are derivatives of Carry Look Ahead Adders.
The Brent-Kung adder [30] is a PPA that requires lesser computational nodes, but it
has the maximum logic depth, resulting in large computation delay. The schematic
of a 32-bit Brent-Kung using black and grey cells adder is illustrated in Fig. 4.8.
The Brent-Kung adder has less wire complexity and a simple architecture, but the
increased logic depth makes it the slowest PPA among the others (refer Table 4.1).

References

1. Pillmeier, M. R. (2002), Barrel shifter design, optimization, and analysis, in 'Lehigh University', Citeseer.
2. Thornton, J. E. (1970), Design of a computer–the control data 6600, Scott Foresman & Co.
3. Palmer, J. F., Ravenel, B. W. and Nave, R. (1982), 'Numeric data processor'. US Patent
   4,338,675.
4. Lim, R. S. (1972), 'A barrel switch design', Computer Design 11(8), 76–78.

5. Neto, H. C. and Vestias, M. P. (2008), Architectural tradeoffs in the design of barrel shifters for reconfigurable computing, in '2008 4th Southern Conference on Programmable Logic', IEEE, pp. 31–36.

6. Pillmeier, M. R., Schulte, M. J. and Walters III, E. G. (2002), Design alternatives for barrel shifters, in 'Advanced Signal Processing Algorithms, Architectures, and Implementations XII', Vol. 4791, International Society for Optics and Photonics, pp. 436–447.

7. Srivastava, P. and Chung, E. (2022), 'An asynchronous bundled-data barrel shifter design that incorporates a deterministic completion detection technique', IEEE Transactions on Circuits and Systems II: Express Briefs 69(3), 1667–1671.

8. Wimer, S., Albeck, A. and Koren, I. (2014), 'A low energy dual-mode adder', Computers & Electrical Engineering 40(5), 1524–1537.

9. Beerel, P. A., Kim, S., Yeh, P.-C. and Kim, K. (1999), Statistically optimized asynchronous barrel shifters for variable length codecs, in 'Proceedings. 1999 International Symposium on Low Power Electronics and Design (Cat. No. 99TH8477)', IEEE, pp. 261–263.

10. Nowick, S. M. (1996), 'Design of a low-latency asynchronous adder using speculative completion', IEE Proceedings-Computers and Digital Techniques 143(5), 301–307.

11. Yih, S.-J., Cheng, M. and Feng, W.-S. (1996), 'Multilevel barrel shifter for CORDIC design', Electronics Letters 32(13), 1178–1179.

12. Mohan, N. and Pragaspathy, I. (2018), 'High performance shifter circuit'. US Patent 9,904,511.

13. Yasuda, T. (2018), 'High speed and low power circuit structure for barrel shifter'. US Patent 9,996,317.

14. Shakya, D. and Agrawal, S. (2018), 'Performance analysis of low power CMOS based 8 bit barrel shifter', International Journal of Research and Analytical Reviews (IJRAR) 5(4), 11–22.

15. Guo, Y., Dee, T. and Tyagi, A. (2018), 'Barrel shifter physical unclonable function based encryption', Cryptography 2(3), 22.

16. Bhardwaj, M. K. K. and Khare, V. (2018), 'Design of shift invert coding using barrel shifter', International Journal of Pure and Applied Mathematics 118(14), 247–252.

17. Kalyanaraman, A. and Manikandan, A. (2018), Adiabatic techniques for energy-efficient barrel shifter design, in 'VLSI Design: Circuits, Systems and Applications', Springer, pp. 27–36.

18. Biswal, L., Bhattacharjee, A., Das, R., Thirunavukarasu, G. and Rahaman, H. (2018), Quantum domain design of Clifford+ t-based bidirectional barrel shifter, in 'International Symposium on VLSI Design and Test', Springer, pp. 606–618.

19. Chaudhuri, S. (2019), 'Selectively combinable directional shifters'. US Patent App. 10/289,382.

20. Bybell, A. J., Frey, B. G. and Gschwind, M. K. (2019), 'Hardware-based pre-page walk virtual address transformation independent of page size utilizing bit shifting based on page size'. US Patent App. 16/283,293.

21. Brown, J. M., Vankayala, V. J., Waldrop, W. C., Mazumder, K., Moon, B. S. and Kandikonda, R. K. (2019), 'Reduced shifter memory system'. US Patent 10,354,717.

22. Bari, S., De, D. and Sarkar, A. (2019), 'Design of low power, high speed 4 bit binary to gray converter with 8 × 4 barrel shifter using nano dimensional MOS transistor for arithmetical, logical and telecommunication circuit and system application', Microsystem Technologies 25(5), 1585–1591.

23. Kirichenko, A. F., Kamkar, M. Y., Walter, J. and Vernik, I. V. (2019), 'ERSQF 8-bit parallel binary shifter for energy-efficient superconducting CPU', IEEE Transactions on Applied Superconductivity 29(5), 1–4.

24. Li, Z., Zhang, G., Zhang, W., Chen, H. and Perkowski, M. (2018), Synthesis of quantum barrel shifters, in 'International Conference on Cloud Computing and Security', Springer, pp. 450–462.

25. Srivastava, P., Chung, E. and Ozana, S. (2020), 'Asynchronous floating-point adders and communication protocols: A survey', Electronics 9(10), 1687.

26. Choi, Y. (2004), Parallel Prefix Adder Design, PhD thesis, The University of Texas at Austin.

27. Chakali, P. and Patnala, M. K. (2013), 'Design of high speed Kogge-Stone based carry select adder', International Journal of Emerging Science and Engineering (IJESE) 1(4), 2319–6378.

28. Gundi, N. D. (2015), Implementation of 32-bit Brent Kung Adder using Complementary Pass Transistor Logic, PhD thesis, Oklahoma State University.
29. Lai, K. K. (2016), Novel Asynchronous Completion Detection For Arithmetic Datapaths, Taylor's University, Malaysia.
30. Brent, R. P. and Kung, H. T. (1982), 'A regular layout for parallel adders', IEEE Transactions on Computers 3, 260–264.
31. Zamhari, N., Voon, P., Kipli, K., Chin, K. L. and Husin, M. H. (2012), Comparison of parallel prefix adder (PPA), in 'Proceedings of the World Congress on Engineering', Vol. 2, pp. 4–6.
32. Xiang, L. M., Mun'im Ahmad Zabidi, M., Awab, A. H. and Ab Rahman, A. A.-H. (2018), VLSI implementation of a fast Kogge-Stone parallel-prefix adder, in 'Journal of Physics: Conference Series', Vol. 1049, IOP Publishing, p. 012077.
33. Lai, K. K., Chung, E. C., Lu, S.-L. L. and Quigley, S. F. (2014), 'Design of a low latency asynchronous adder using early completion detection', Journal of Engineering Science and Technology 9(6), 755–772.
34. Ramanathan, P. and Vanathi, P. (2009), 'Hybrid prefix adder architecture for minimizing the power delay product', International Journal of Electrical and Computer Engineering 4, 9.
35. Kogge, P. M. and Stone, H. S. (1973), 'A parallel algorithm for the efficient solution of a general class of recurrence equations', IEEE transactions on Computers 100(8), 786–793.
36. Sklansky, J. (1960), 'An evaluation of several two-summand binary adders', IRE Transactions on Electronic Computers 2, 213–226.
37. Han, T. and Carlson, D. A. (1987), Fast area-efficient VLSI adders, in '1987 IEEE 8th Symposium on Computer Arithmetic (ARITH)', IEEE, pp. 49–56.
38. Ladner, R. E. and Fischer, M. J. (1980), 'Parallel prefix computation', Journal of the ACM (JACM) 27(4), 831–838.
39. Nowick, S. M., Yun, K. Y., Beerel, P. A. and Dooply, A. E. (1997), Speculative completion for the design of high-performance asynchronous dynamic adders, in 'Proceedings Third International Symposium on Advanced Research in Asynchronous Circuits and Systems', IEEE, pp. 210–223.
40. Thakur, G., Sohal, H. and Jain, S. (2021), 'A novel ASIC-based variable latency speculative parallel prefix adder for image processing application', Circuits, Systems, and Signal Processing 40, 5682–5704.
41. Esposito, D., De Caro, D. and Strollo, A. G. M. (2016), 'Variable latency speculative parallel prefix adders for unsigned and signed operands', IEEE Transactions on Circuits and Systems I: Regular Papers 63(8), 1200–1209.
42. Cilardo, A. (2015), Variable-latency signed addition on FPGAs, in '2015 25th International Conference on Field Programmable Logic and Applications (FPL)', IEEE, pp. 1–6.
43. Esposito, D., De Caro, D., Napoli, E., Petra, N. and Strollo, A. G. M. (2015), 'Variable latency speculative Han-Carlson adder', IEEE Transactions on Circuits and Systems I: Regular Papers 62(5), 1353–1361.
44. Hand, D., Cheng, B., Breuer, M. and Beerel, P. A. (2015), 'Blade–a timing violation resilient asynchronous design template'.

Chapter 5
Generic Architecture of Deterministic Completion Detection Scheme

5.1 Design Methodology

The main objective of this book is to develop a generic architecture of a deterministic completion detection scheme for asynchronous circuits using bundled data protocol. The literature review was conducted to analyse the recent trends, methodologies, and challenges associated with asynchronous design style, and a research gap in the existing literature was identified. The design methodology used to achieve the objectives of this book contains three major phases: (i) Preliminary Stage, (ii) Design and Development Stage, and (iii) Application Stage.

5.1.1 Preliminary Stage

In the first stage, the existing literature on asynchronous circuit design was reviewed. The communication protocols required to synchronise the process in asynchronous circuits due to absence of a clock signal were thoroughly investigated. Each communication protocol requires a completion detection scheme to indicate **when** an ongoing computation process has completed. Existing completion detection schemes were compared, indicating that when combined with an optimal completion detection scheme, the bundled data protocol exhibits promising results.

The speculative completion detection scheme with an abort network is commonly used with bundled data protocol, which improves the speed of a computation block. This design is further modified for asynchronous adders [1] by omitting the abort network, as implementing the abort logic is complex, rendering the speculative method inapplicable for fast adders. While Lai's work was limited to asynchronous adders, it paves the way for an asynchronous circuit design scheme using bundled data protocol that can deterministically determine the completion of an event. Hence, the research gap was identified and research objectives were defined in the first phase through a thorough literature review. The challenges inherent in achieving these objectives were also identified in the first phase. The following are the key points of the preliminary stage:

- Identify the need for compact, high-performance electronic devices with minimal power consumption.
- Study the history of asynchronous design methodologies and their implementation in a variety of scientific and engineering applications.
- Examine various protocols for communicating between two asynchronous blocks and determining data validity to indicate the completion of an ongoing event.
- Review the existing completion detection schemes.

5.1.2   Design and Development Stage

The second stage is focused on the theoretical development of a design in order to accomplish the objectives defined in the first stage. A generic model for a deterministic completion detection scheme was developed for bundled data circuits that can be applicable for fast adders and other high-speed computation blocks, as it omits the complex abort network. A necessary condition was defined for implementing a computation block using the proposed model, and Pallavi's generic function G_p was defined. The proposed model was then theoretically validated using an adder and a barrel shifter, as addition and shift are two fundamental operations of arithmetical and logical units, respectively. The key contributions of the design and development stage are:

- Develop a generic approach for implementing a deterministic completion detection scheme for bundled data circuits.
- Design the architecture of a single-precision asynchronous bundled data barrel shifter to validate the proposed generic architecture.

5.1.3   Application Stage

In the third and final stages, the research design developed in the second stage was experimentally validated on binary adder and barrel shifter. All barrel shifter and adder designs addressed in this book are coded in Verilog HDL. Additionally, delays at the gate level are incorporated into the codes to perform timed simulation and to determine the propagation delay of the circuit. The computation delays of the adder and shifter designed with the proposed scheme were compared with their synchronous counterparts. Throughout this research, Xilinx Vivado ® Design Suite was used to conduct all experiments.

5.2 Deterministic Completion Detection Scheme

The main objective of this book is to develop a deterministic completion detection scheme for bundled data circuits. The existing completion detection schemes for bundled data circuits demonstrated the potential for performance improvement, as discussed in Chap. 3. It can be concluded from the literature review that there are still various aspects of bundled data protocol that need to be investigated [5, 6]. The speculative completion detection scheme demonstrated the performance boost of asynchronous adder ranging from 5% to 30% over a comparable synchronous implementation [2], but the abort network used in this approach adversely affected the overall performance of the adder and thus cannot be applied for faster adders [1]. A deterministic completion detection circuit for asynchronous adders was developed such that it eliminates the need for an abort network, and the adder delay is reduced by 8% to 10% [1, 3]; yet this approach was limited to designing an asynchronous adder.

Previous research can only be viewed as a first step towards the development of bundled data circuits with a completion detection scheme. This research aims to explore the asynchronous design style for bundled data protocol and to develop a generic model of deterministic completion detection scheme for asynchronous circuits using bundled data protocol. The generic architecture of a single-rail bundled data circuit with a deterministic completion detection scheme is shown in Fig. 5.1. The deterministic completion detection scheme consists of a Deterministic Completion Detection Circuit (DCDC) and a Delay Generating Unit (DGU).

The DGU contains multiple delay models in order to replicate the propagation delays of active signal paths corresponding to all possible input combinations of

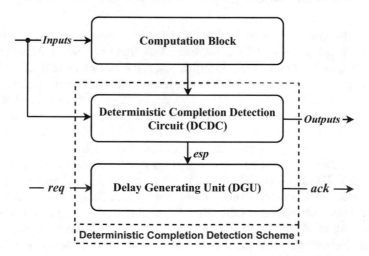

Fig. 5.1 Generic architecture of a bundled data channel with deterministic completion detection scheme

ζ. DCDC is responsible for computing the propagation delay of the unique active signal path corresponding to an ongoing computation and generates an *enable shorter path (esp)* signal to select a corresponding delay model from DGU.

> Unlike the speculative scheme, where *late* signal is used to abort the default early completion path in case of late completion detection, the *esp* signal in the deterministic scheme assists the DGU to activate a unique delay path corresponding to the active signal path of ζ associated with the ongoing computation.

The DCDC comprises two blocks: Input Dependent Selector (IDS) and Output Selection Stage (OSS). The IDS assists the OSS in selecting the output of the computation block ζ after the propagation delay associated with the active signal path required by the computation process. OSS selects the output data from the final active stage of ζ after which the value of output data will not change. IDS is responsible for deciphering the active stages of ζ associated with the input data and selecting the corresponding delay path from the DGU. Delay models in DGU are designed to be slightly slower than the actual computation block, and the longest path available in the delay model has the delay value slightly greater than the worst-case delay of the computation block (refer Sect. 5.2.2). The next few sections demonstrate the working of DCDC and DGU in detail.

5.2.1 Deterministic Completion Detection Circuit (DCDC)

DCDC is designed to choose the optimal delay path from DGU corresponding to the ongoing computation. DCDC consists of an Input Dependent Selector (IDS) and an Output Selection Stage (OSS) to assist the circuit in selecting the final output of ζ, as shown in Fig. 5.2. IDS operates parallel to the computation block and assists the OSS in detecting the final active stage of ζ associated with the current input combination, after which the state of output remains unchanged. The output of this final stage is then captured by OSS, which represents the final output of ζ corresponding to the current input combination. IDS also determines the propagation delay required by active signal path of ζ associated with the ongoing computation. IDS is responsible for generating *esp* signals in order to activate the delay model in DGU corresponding to the propagation delay of active signal path associated with the ongoing computation. Depending on the computation properties of ζ, the output of IDS can be interpreted either as combinations of input values and/or as derivative of these input combinations. Any digital computation block ζ, such as an adder, a shifter, or even a processor, is designed with a finite data width n. The following axiom is defined in this book stating that the number of input

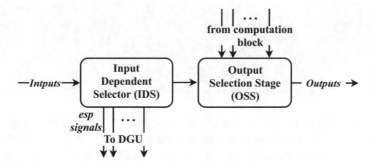

Fig. 5.2 Deterministic completion detection circuit

combinations for an n-bit ζ is computed in terms of the finite value n, implying that it can only process a finite number of input combinations.

Axiom 5.1

For any computation block ζ that connects one or more inputs to one output, there exists a finite combination of input values x_i such that $X = \{x_1, x_2, \ldots, x_\epsilon\}$ denotes the set of all possible input values for ζ and $|X| = \epsilon$.

The variable ϵ represents the maximum possible input values/combinations for ζ, and it can be calculated in terms of the data width n. The propagation delays required to process these ϵ input combinations are determined by their associated active signal paths, which are chosen from the set of all possible unique critical paths of ζ. Thus, for each input x_i, there exists a corresponding critical path $\mathcal{P}_j$ from input to output, where the critical path is the path with the longest delay. The following definition is introduced in this book for the critical path corresponding to the input data:

Definition 5.1 The set $\mathcal{P} = \{\mathcal{P}_1, \mathcal{P}_2, \ldots, \mathcal{P}_\gamma\}$ is defined as the set of all possible critical paths: $\forall\, x_i \in X\ \exists\, \mathcal{P}_j \in P$, where $\mathcal{P}_j$ is the critical path corresponding to the input data x_i, and $|\mathcal{P}| = \gamma$.

The variable γ denotes the total number of unique critical paths—again critical path being the slowest path connecting input to output. Each critical path from the set $\mathcal{P}$ is associated with a unique propagation delay, which is defined in this book as follows:

Definition 5.2 The set $\Delta = \{\delta_1, \delta_2, \ldots, \delta_\gamma\}$ is defined as the set of propagation delays corresponding to all critical paths. In other words, $\forall \mathcal{P}_i \in \mathcal{P}$ $\exists \delta_i \in \Delta$, where δ_i is the propagation delay associated with the critical path $\mathcal{P}_i$.

As stated in Axiom 5.1, ϵ denotes the total number of possible input combinations for ζ. These ϵ input combinations utilise γ paths of ζ, as mentioned in Definition 5.1. The relation between ϵ and γ is established in this book as given in the following axiom:

Axiom 5.2

For any digital computation block ζ, there exists a finite number of all possible critical paths γ such that $\gamma \leq \epsilon$.

As stated in Axiom 5.1, $|X| = \epsilon \in \mathbb{Z}_+$ is finite because the input to a digital computation block is always in binary form (either logic 0 or logic 1), resulting in a finite number of all possible input combinations for an n-bit digital computation block. If each input from the finite number of input combinations utilises a unique critical path, then $|\mathcal{P}| = |X|$. However, this is not a standard scenario and a unique critical path is generally utilised by more than one input combination in a digital computation block, providing $|\mathcal{P}| < |X|$. Since $|\mathcal{P}| = \gamma$ and $|X| = \epsilon$, combining the above conditions provides $\gamma \leq \epsilon$.

From Axiom 5.1, it can be concluded that since ϵ is finite, γ is also finite, implying that there is a known finite number of delays associated with the computation block ζ. Now, in order to develop a deterministic completion detection scheme for an asynchronous digital computation block ζ, the following condition must be satisfied.

Condition C1

For each $x_i \in X \exists \mathcal{G}_p: \mathcal{G}_p(x_i) \mapsto \mathcal{P}_j$, where $\mathcal{P}_j \in \mathcal{P}$ is the critical path associated with x_i. The function $\mathcal{G}_p$ is referred to as Pallavi's generic function in this book.

In other words, the condition C1 defined in this book for deterministic completion detection scheme states that for a digital computation block ζ, there exists a known function $\mathcal{G}_p$ that can determine the critical path associated with each input combination from the input value itself and/or its derivative. Figure 5.3 illustrates the implementation of $\mathcal{G}_p$ for a computation block ζ with an OSS and DGU. Accordingly, with a known function $\mathcal{G}_p$ for a given computation block, one can determine the delay it requires for its output to be ready, i.e., completion of the

Fig. 5.3 Pallavi's generic function $\mathcal{G}_p$

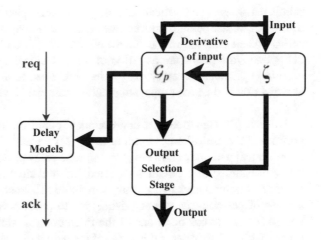

computation, as soon as the input arrives at ζ. It is worth noting that the block in Fig. 5.3 that implements the generic function $\mathcal{G}_p$ is identical to the IDS block in Fig. 5.2. Hence, implementing the generic function will result in the implementation of the IDS block. Throughout the book, IDS and $\mathcal{G}_p$ are used interchangeably.

5.2.2 Delay Generating Unit (DGU)

DGU contains multiple delay models, each replicating the propagation delay of the active critical path associated with the current input combination. When DGU receives the *req* signal, it is responsible for generating *ack* signal after τ_i delay, τ_i being the propagation delay of the delay model associated with the active signal path $\mathcal{P}_i$ corresponding to current input combination x_j. The propagation delay of the DGU is represented by the set $T = \{\tau_1, \tau_2, \ldots, \tau_\gamma\}$ corresponding to the delay set $\Delta = \{\delta_1, \delta_2, \ldots, \delta_\gamma\}$ of the actual computation block ζ. Two different variables $\delta \in \Delta$ and $\tau \in T$ are used to represent the delay of ζ and DGU, respectively, in order to make it easier to distinguish the delay of these two blocks. The following definition is introduced in this book for delay models in DGU:

Definition 5.3 The set $T = \{\tau_1, \tau_2, \ldots, \tau_\gamma\}$ is defined as the set of propagation delays of delay models in DGU corresponding to all critical paths of ζ, i.e., $\forall \mathcal{P}_i \in \mathcal{P} \, \exists \, \tau_i \in T$ and $|T| = |\mathcal{P}|$.

In other words, for every critical path of ζ, there exists a corresponding delay model in DGU. It is essential to emphasise that these delay models must precisely mimic the delays of all critical paths of ζ. An overly pessimistic delay model results in the *ack* signal being produced later than the propagation delay required by the active critical path to process the event, whereas an underestimation of the actual delay leads to a premature assertion of the *ack* signal. As a result, the logic gates used to implement the delay models in DGU need to at best perfectly match and at worst slower than the propagation delays associated with the active critical paths of ζ.

Nowick [2] suggested that delay models could be implemented by extracting portions of the critical paths in order to perfectly match the delays of delay models to the critical paths of ζ; however, this approach may provide an erroneous outcome. It is well established in circuit design that different circuit corners can have different effects on circuit delays [4], as shown in Table 5.1. Each element of ζ may exhibit a range of performance characteristics due to process variation, temperature, etc. i.e., each element can be in one of the three corners: slow, typical, or fast. If each row in Table 5.1 represents the speed of the actual digital computation block ζ and each column represents the speed of the corresponding delay model, the preferred design arrangement is that the delay models in DGU are taken slower than the propagation delay associated with active critical path of ζ, represented by green cells. The red cells clearly depict the forbidden implementation of DGU, as DGU cannot have faster delay than ζ. The condition that the speed of the critical path in the actual circuit matches the speed of the delay model is too stringent, as ζ may have physical constraints such as complex wiring, longer routes, higher output capacitance, higher fan-outs, etc., which can affect the propagation delay of the active path of ζ. Therefore, the gray cells in Table 5.1 representing equal delays of ζ and DGU are not preferred.

It is critical to design the delay models that operate slower than the corresponding active signal paths of ζ under all physical conditions, in order to detect the valid output data. The necessary condition for error-free completion detection and for generating the *ack* signal after the propagation delay associated with active critical path of ζ is defined in this book as follows:

Condition C2

For any computation block ζ, there exists $\tau_i \in T$ corresponding to $\delta_i \in \Delta : \tau_i > \delta_i$ under all physical conditions and for all possible input combinations.

Table 5.1 Relation between delays of DGU and ζ

DGU	ζ		
	Slow	Typical	Fast
Slow			
Typical			
Fast			

In other words, the propagation delay of the delay models in DGU must be slower than the propagation delay associated with the active critical path of ζ under all physical conditions and for all possible input combinations. The above condition allows the DGU to ensure that the ack signal is never asserted before the propagation delay of active critical path of ζ. The delay models in DGU must be designed to mimic the critical paths of the computation block ζ, and the optimal solution for creating these matching delay models for DGU is to use the same logic elements that implement the signal paths of ζ. The challenge is to determine how to make the delay model run slower than the actual digital computation block ζ, as mentioned in Condition C2. There are two possibilities to implement a slower design for DGU

1. The gates used in the delay model should have more inputs than the actual gates of ζ and/or
2. The gates used in the delay model should be of a weaker drive strength than the actual gates of ζ.

This is to ensure that the propagation delay through the delay model is always larger than the propagation delay of active critical path of ζ. In practice, however, there may be other timing constraints between the ack signal and the validity of the output data that needs to be addressed, depending on how the ack signal is used.

5.3 Chapter Summary

This chapter outlines the research framework and methodology that was employed to conduct this study. This chapter details each of the three activities required by this framework, followed by a discussion about research methodology for this study. This chapter discusses a generic approach for deterministically determining the completion of an event in asynchronous circuits utilising bundled data protocol. The key components required to develop a deterministic completion detection scheme were presented. The conditions to develop a deterministic completion detection scheme for error-free completion detection were established under which the generic design holds true.

References

1. Lai, K. K. (2016), Novel Asynchronous Completion Detection For Arithmetic Datapaths, Taylor's University, Malaysia.
2. Nowick, S. M. (1996), 'Design of a low-latency asynchronous adder using speculative completion', IEE Proceedings-Computers and Digital Techniques 143(5), 301–307.
3. Koes, D., Chelcea, T., Onyeama, C. and Goldstein, S. C. (2005), Adding faster with application specific early termination, Technical report, Carnegie-Mellon Univ Pittsburgh PA publisher of Computer Science.

4. Bhattacharjee, P., Bhattacharyya, B. and Majumder, A. (2021), 'A vector-controlled variable delay circuit to develop near-symmetric output rise/fall time', Circuits, Systems, and Signal Processing 40, 1569–1588.
5. Srivastava, P., Chung, E. and Ozana, S. (2020), 'Asynchronous floating-point adders and communication protocols: A survey', Electronics 9(10), 1687.
6. Srivastava, P. and Chung, E. (2022), 'An asynchronous bundled-data barrel shifter design that incorporates a deterministic completion detection technique', IEEE Transactions on Circuits and Systems II: Express Briefs 69(3), 1667–1671.

Chapter 6
Architecture Optimisation Using Deterministic Completion Detection

6.1 Asynchronous Bundled Data Barrel Shifter

The architecture of a single-precision right-shift ABBS using generic architecture of deterministic completion detection scheme is explored in this section, as illustrated in Fig. 6.1. While this study focuses on implementing a 32-bit ABBS with a right-shift operation, the concept stands true for any n-bit right-/left-shift ABBS.

Notation
The architecture of an 8-bit right-shift CBS is explained in Chap. 4, and a 32-bit CBS can be structured in a similar way, providing a shift range from 0 to 31 bits. As mentioned in Definition 5.1, the total number of all possible input combinations is given as $\epsilon = \epsilon_d \times \epsilon_s$, where (i) ϵ_d represents the total number of all possible input combinations for input data to be shifted, $\epsilon_d = 2^{32} = 4,294,967,296$, and (ii) ϵ_s represents the total number of all possible input combinations for shift amount $\epsilon_s = 2^5 = 32$. Since only the input s (shift amount) is responsible to determine the active stages of the CBS, the average-case delay of the shifter is determined by the total number of all possible input combinations for shift amount ϵ_s. The set of critical paths $\mathcal{P} = \{\mathcal{P}_1, \mathcal{P}_2, \ldots, \mathcal{P}_6\}$ with associated propagation delays $\Delta = \{\delta_1, \delta_2, \ldots, \delta_6\}$ and the delay models in DGU is designed with propagation delays $T = \{\tau_1, \tau_2, \ldots, \tau_6\}$. The total number of paths $\gamma = 6$, 5 paths associated with 5 stages of CBS, and one special path for shift amount $s = 0$, represented by *path-s* and *path-0* in Fig. 6.2, respectively.

The shift amount is capable of determining the number of active stages of shifter, as soon as the input arrives at the shifter. Condition C1 established in Chap. 5 is satisfied for the shifter as $G_p(s) \mapsto \mathcal{P}_j$, i.e., active critical path of the shifter can be determined by the shift amount, and hence a deterministic completion scheme can be developed for the CBS. Figure 6.2 illustrates the block diagram

© The Author(s), under exclusive license to Springer Nature Switzerland AG 2022
P. Srivastava, *Completion Detection in Asynchronous Circuits*,
https://doi.org/10.1007/978-3-031-18397-3_6

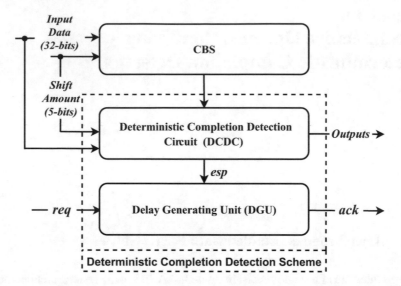

Fig. 6.1 ABBS design using generic architecture of deterministic completion detection scheme

of a single-precision right-shift ABBS. ABBS consists of a single-precision right-shift Conventional Barrel Shifter (CBS) and a deterministic completion detection scheme, as discussed in the following sections.

6.1.1 Single-Precision Conventional Barrel Shifter

A CBS can shift an n-bit data by s-bits in a single operation for $0 \le s \le n\text{-}1$, depending upon the value of select line S_i of the shifter. The propagation delay of an n-bit CBS can be given as

$$\delta_{CBS} = \lambda \times \delta_{stage}, \tag{6.1}$$

where δ_{stage} is the propagation delay of a single multiplexer stage of a CBS, and λ denotes the total number of multiplexer stages. The propagation delay of a digital circuit is defined as the delay required for the data to travel from input to output through all the intermediate stages of the digital circuit.

The architecture of an 8-bit right-shift CBS is explained in Chap. 4, and a 32-bit CBS can be structured in a similar way, providing a shift range from 0 to 31 bits. A single-precision CBS can shift a 32-bit data by s-bits, where $0 \le s \le 31$, with the total number of shift stages $\lambda = \log_2 32 = 5$, as shown in Figs. 6.3 and 6.4.

A CBS stage i can shift the data by 2^i bits for $S_i = 1$, else it will pass the data to its output without shifting for $S_i = 0$. The shift amount s can be expressed in terms

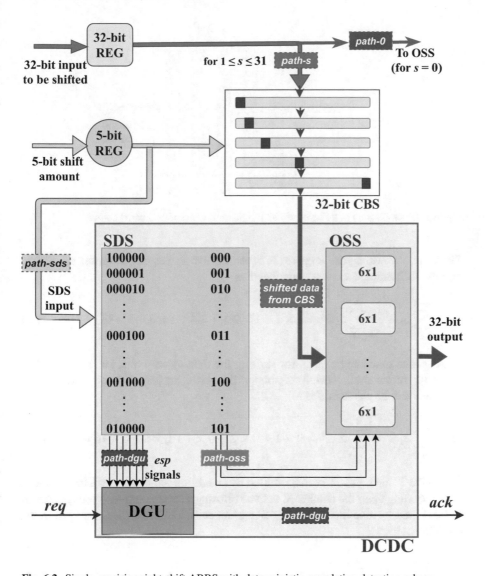

Fig. 6.2 Single-precision right-shift ABBS with deterministic completion detection scheme

of the select lines of the CBS stages, as given in Eq. 4.1, and hence the number of active stages of CBS depends upon s. For instance, the best-case scenario is when only a single stage S_0 is active. The shift amount s can be calculated using Eq. 4.1 as

$$s = \sum_{i=0}^{\lambda-1} S_i.2^i = \sum_{i=0}^{4} S_i.2^i = 1 \times 1 + 0 \times 2 + 0 \times 4 + 0 \times 8 + 0 \times 16 = 1. \quad (6.2)$$

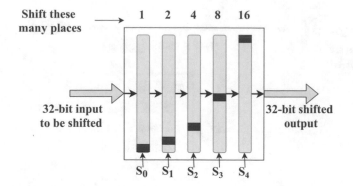

Fig. 6.3 Block diagram of a single-precision right-shift conventional barrel shifter

Similarly, the worst-case scenario is when all five stages of the shifter are active, and the shift amount s can be calculated as

$$s = \sum_{i=0}^{\lambda-1} S_i.2^i = \sum_{i=0}^{4} S_i.2^i = 1 \times 1 + 1 \times 2 + 1 \times 4 + 1 \times 8 + 1 \times 16 = 31. \quad (6.3)$$

A special case can be taken for shifting the data by zero bits, i.e., the data need not to be shifted at all. Therefore, none of the shifter stages are active, and the shift amount s can be calculated as

$$s = \sum_{i=0}^{\lambda-1} S_i.2^i = \sum_{i=0}^{4} S_i.2^i = 0 \times 1 + 0 \times 2 + 0 \times 4 + 0 \times 8 + 0 \times 16 = 0. \quad (6.4)$$

Hence the 32-bit input data can be shifted by an arbitrary amount s such that $0 \le s \le 31$, depending upon the number of active shift stages. For example, if the data needs to be shifted by nine bits, then only S_0 and S_3 stages will be active as shown in the equation below:

$$s = \sum_{i=0}^{\lambda-1} S_i.2^i = \sum_{i=0}^{4} S_i.2^i = 1 \times 1 + 0 \times 2 + 0 \times 4 + 1 \times 8 + 0 \times 16 = 9. \quad (6.5)$$

It is evident from Eqs. 6.2, 6.4, and 6.5 that all five stages are not active for these shift amounts, as the number of active shift stages depends upon the value of s. The critical path utilising all five stages of a single-precision right-shift CBS consumes the maximum delay, as illustrated in Fig. 6.4. The worst-case propagation delay of a single-precision CBS can be calculated from Eq. 6.1 as

$$\delta_{CBS} = \lambda \times \delta_{stage} = 5 \times \delta_{stage}. \quad (6.6)$$

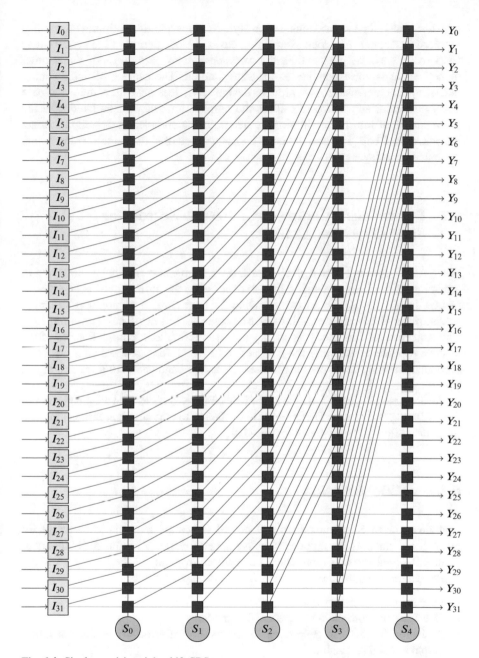

Fig. 6.4 Single-precision right-shift CBS

The clock period of a synchronous CBS is determined by the worst-case delay associated with its critical path, assuming CBS is the only logic connecting two registers. Since the synchronous design style imposes additional timing constraints required for setup time, clock skew, jitter, etc., the clock period of the synchronous CBS will always be greater than the worst-case delay, regardless of the number of active stages corresponding to the shift amount. The same CBS architecture, when implemented using asynchronous design style, does not follow the worst-case delay constraint and therefore utilises the propagation delays associated with the active stages of the shifter circuit. The deterministic completion detection scheme introduced in Chap. 5 is used to detect the completion of shift operation after actual computation delay, as discussed in the next section.

6.2   Deterministic Completion Detection Scheme for Single-Precision ABBS

Since ABBS does not have a clock signal, there is no worst-case delay constraint to assist the shifter in commencing the next shift after the current shift is completed. Hence, to aid the ABBS in commencing its next shift, a completion detection scheme is required for determining when an ongoing shifting process has completed. The objective of this study is to develop a deterministic completion detection scheme capable of detecting the completion of the shifting process after an actual computation delay corresponding to the shift amount and generating the ack signal accordingly.

As discussed in Chap. 5, the deterministic completion detection scheme is implemented using two blocks: DCDC and DGU. The following sections demonstrate the implementation of these blocks for an ABBS.

6.2.1   DCDC for ABBS

Equations 6.2 to 6.5 demonstrate that not all stages are required to be active for all shift amounts, as the number of active stage(s) varies with the shift amount s. Hence, the final shifted output could be captured from any of the five stages of the shifter depending upon the value of s. DCDC is responsible for

- Determining the active shifter stage(s)
- Detecting the completion of the shifting process
- Assisting the DGU in generating the ack signal after ensuring that the shifter output is valid and stable

As mentioned in Chap. 5, DCDC comprises an Output Selection Stage (OSS) and an Input Dependent Selector (IDS) [1]. The DCDC for Single-Precision Right-Shift ABBS also consists of an OSS, and the block IDS is renamed as Shift Dependent Selector (SDS), as shown in Fig. 6.5, and described in the next few sections.

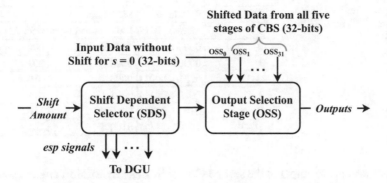

Fig. 6.5 Deterministic completion detection circuit for single-precision right-shift ABBS

6.2.1.1   Output Selection Stage (OSS)

OSS is implemented using 32 6×1 multiplexers, one for each bit of the 32-bit input data. These multiplexers assist the ABBS in selecting the shifted output from one of the five stages corresponding to the shift amount $1 \leq s \leq 31$, as shown by the signal path *path-s* in Fig. 6.2. The signal path *path-0* represents a special case of shifter for shift amount $s = 0$, as it bypasses the CBS and directs the 32-bit input data to the OSS stage without shifting. The final output of the ABBS is captured from the OSS stage with the help of its 3-bit select lines, once the active shifter stage(s) corresponding to the shift amount have been identified.

6.2.1.2   Shift Dependent Selector (SDS)

SDS implements Pallavi's generic function $\mathcal{G}_p$ for ABBS. SDS is responsible for deciphering the number of active shifter stage(s) as the generic function $\mathcal{G}_p$ maps the input shift amount into the corresponding delay path from the DGU. An SDS is configured with 32 memory locations, each of which is 9-bits wide for a single-precision ABBS, as specified in Table 6.1. The 5-bit shift amount is given to the SDS input, which acts as a unique address of these memory locations ($2^5 = 32$) corresponding to the shift amount. The term SDS output refers to the 9-bit data stored in these memory locations.

The first three least significant bits of SDS output are connected to the select lines of OSS, as shown by the subpath *path-oss* of the signal path *path-sds*. This allows the OSS to select the earliest internal signal within the CBS known to carry the valid output value. The OSS is programmed to select one of its inputs such that

1. OSS selects the 32-bit input data without any shift for shift amount $s = 0$ and OSS select lines $= 000$.
2. OSS selects the output of the first CBS shift stage S_0 for shift amount $s = 1$ and OSS select lines $= 001$.

Table 6.1 Shift dependent selector (SDS) for a 32-bit ABBS

Shift amount	SDS input	SDS output	
s	$(S_4 S_3 S_2 S_1 S_0)$	Delay model input	OSS select line
0	00000	100000	000
1	00001	000001	001
2–3	00010–00011	000010	010
4–7	00100–00111	000100	011
8–15	01000–01111	001000	100
16–31	10000–11111	010000	101

3. OSS selects the output of the second CBS shift stage S_1 for shift amount $s = 2, 3$ and OSS select lines $= 010$.
4. OSS selects the output of the third CBS shift stage S_2 for shift amount $s = 4$ to 7 and OSS select lines $= 011$.
5. OSS selects the for output of the fourth CBS shift stage S_3 for shift amount $s = 8$ to 15 and OSS select lines $= 100$.
6. OSS selects the for output of the fifth CBS shift stage S_5 for shift amount $s = 16$ to 31 and OSS select lines $= 101$.

The remaining 6-bits of SDS output are used to generate the esp signals for the DGU in order to generate the ack signal after the propagation delay associated with the shift amount. The signal paths for this case are illustrated in Fig. 6.2 as the subpath $path$-dgu of the signal path $path$-sds.

6.2.2   DGU for ABBS

DGU is responsible for asserting the ack signal after propagation delay associated with the active critical path of the shifter. As mentioned in Chap. 5, DGU contains multiple delay models that are designed to replicate the propagation delays of the various signal paths present in the shifter. The most common way of implementing a delay model is by utilising a chain of inverters, and however the optimal solution for creating matching delay models for DGU is to use the same logic elements that implement the signal paths of the shifter. Since the CBS stages and stage OSS are built using multiplexers, the delay models in DGU are implemented using basic gates designed to function like multiplexers, as shown in Fig. 6.6.

The delay of shifter and its corresponding delay model are represented by the variables δ and τ, respectively. Therefore, τ_{stage} signifies the delay of the delay model associated with the CBS stage delay δ_{stage}, whereas τ_{oss} specifies the delay of the delay model that corresponds to the CBS stage delay δ_{oss}. The delay model corresponding to a single CBS stage is built using two AND gates, one OR gate and one inverter, similar to a 2×1 multiplexer. Likewise, the delay model corresponding to OSS stage is built using basic gates configured to behave as 6×1 multiplexers. As stated in Condition C2 in Chap. 5, the propagation delay of the DGU must be slower

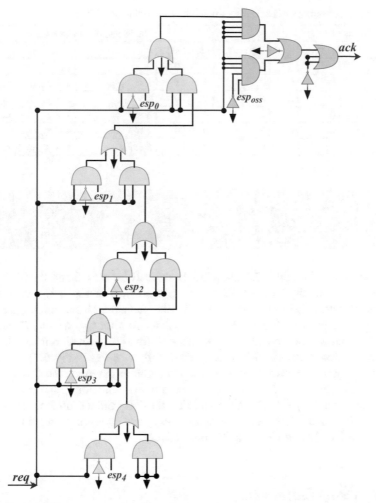

Fig. 6.6 Delay models in delay generating unit (DGU) for single-precision asynchronous bundled data barrel shifter (ABBS)

than the propagation delay of the shifter, i.e., $\tau_i > \delta_i$; therefore, the gates used in the delay model have more inputs than the actual gates in the multiplexers in CBS.

SDS generates esp signals in order to select the appropriate delay model from DGU corresponding to the shift amount (refer path $path\text{-}dgu$ in Fig. 6.2). Table 6.2 shows the value of esp signals corresponding to the shift amount and the delay associated with it. Signals esp_0 to esp_4 select the delay models

(continued)

Table 6.2 Input settings of DGU for a single-precision right-shift ABBS

Shift amount	DGU input						CBS stage delay	DGU delay
s	esp_{oss}	esp_4	esp_3	esp_2	esp_1	esp_0	δ	τ
0	1	0	0	0	0	0	0	τ_{oss}
1	0	0	0	0	0	1	$1 \times \delta_{stage}$	$1 \times \tau_{stage} + \tau_{oss}$
2–3	0	0	0	0	1	0	$2 \times \delta_{stage}$	$2 \times \tau_{stage} + \tau_{oss}$
4–7	0	0	0	1	0	0	$3 \times \delta_{stage}$	$3 \times \tau_{stage} + \tau_{oss}$
8–15	0	0	1	0	0	0	$4 \times \delta_{stage}$	$4 \times \tau_{stage} + \tau_{oss}$
16–31	0	1	0	0	0	0	$5 \times \delta_{stage}$	$5 \times \tau_{stage} + \tau_{oss}$

corresponding to the CBS stages 0 to 4, and signal esp_{oss} selects the delay model associated with the shift amount $s = 0$.

As explained in Sect. 6.2.1, the CBS is skipped for $s = 0$ and the 32-bit input data is given directly to the OSS stage (refer *path*-0 in Fig. 6.2), which can be selected by setting OSS select lines as 000 (refer Table 6.1); hence only signal esp_{oss} will be active for $s = 0$ and *ack* signal is asserted after τ_{oss} delay. Likewise, the shifted output will be available at CBS stage S_0 for $s = 1$, which can be selected by setting OSS select lines as 001 (refer *path-s* and *path-oss* in Fig. 6.2); hence only signal esp_0 will be active for $s = 1$ and *ack* signal is asserted after ($\tau_{stage} + \tau_{oss}$) delay. Similarly, stages S_1, S_0, and OSS are involved in shifting the data by 2 and 3-bits, and signal esp_1 will be active to select the delay model corresponding to shift amount $s = 2$ & 3. Likewise, the settings of esp_3 and esp_4 are arranged for the shift values $s = 8$ to 15 and $s = 16$ to 31, respectively.

6.2.3 Delay Calculation

The shifter design implemented using the deterministic completion detection scheme assists the shifter to utilise the propagation delay associated with active critical path corresponding to the shift amount. This design style allows the shifter to operate on an average-case delay, as discussed in Chap. 5. The average-case delay of the shifter can be calculated theoretically using Eq. 2.1 as

Table 6.3 Path repetitivity in a single-precision right-shift ABBS

Shift amount	ABBS delay	Repetitivity
0	$\delta_1 = \delta oss$	1
1	$\delta_2 = (1 \times \delta_{stage} + \delta_{oss})$	1
2–3	$\delta_3 = (2 \times \delta_{stage} + \delta_{oss})$	2
4–7	$\delta_4 = (3 \times \delta_{stage} + \delta_{oss})$	4
8–15	$\delta_5 = (4 \times \delta_{stage} + \delta_{oss})$	8
16–31	$\delta_6 = (5 \times \delta_{stage} + \delta_{oss})$	16

$$\delta_{ABBS} = \sum_{i=1}^{\gamma_c} \left(\frac{\delta_i . r_i}{\epsilon_s} \right) = \sum_{i=1}^{6} \left(\frac{\delta_i . r_i}{32} \right), \tag{6.7}$$

where the total number of input combinations for shift amount is $\epsilon_s = 32$ for a single-precision ABBS and $\gamma_c = 6$ paths are available: *path-s* has 5 different paths, one for each stage output, and *path-0* is for $s = 0$. Table 6.3 shows the repetitivity of a signal path corresponding to a shift amount and the ABBS delay associated with it.

The average delay of the ABBS can be further calculated as

$$\delta_{ABBS} = \delta_1 \times \left(\frac{1}{32} \right) + \delta_2 \times \left(\frac{1}{32} \right) + \delta_3 \times \left(\frac{2}{32} \right) + \delta_4 \times \left(\frac{4}{32} \right) + \delta_5 \times \left(\frac{8}{32} \right) + \delta_6 \times \left(\frac{16}{32} \right)$$

$$= \delta_1 \times \left(\frac{1}{32} \right) + \delta_2 \times \left(\frac{1}{32} \right) + \delta_3 \times \left(\frac{1}{16} \right) + \delta_4 \times \left(\frac{1}{8} \right) + \delta_5 \times \left(\frac{1}{4} \right) + \delta_6 \times \left(\frac{1}{2} \right). \tag{6.8}$$

As shown in Eq. 6.8, the propagation delay of ABBS is calculated using the propagation delay associated with the active critical path corresponding to the shift amount. Therefore, ABBS operates on an average-case delay, in contrast to CBS, which always operates on the worst-case delay (refer Eq. 6.6) irrespective to the shift amount. In real-world applications, the distribution of shift values is not uniform [2], and the shifter average-case delay can be further optimised for floating-point adders (FPAs), as discussed in the next chapter.

6.3 Asynchronous Bundled Data Binary Adder

The architecture of a single-precision ABBA [3] using the generic architecture of deterministic completion detection scheme is explored in this section, as illustrated

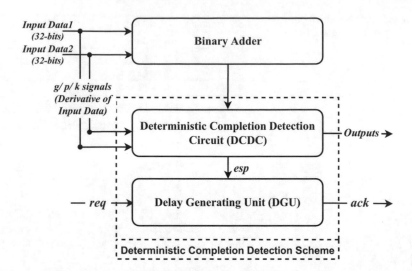

Fig. 6.7 ABBA design using generic architecture of deterministic completion detection scheme

in Fig. 6.7. While this study focuses on implementing a 32-bit ABBA, the concept stands true for any n-bit ABBA.

Notation
As stated in Definition 5.1, the total number of possible input combinations for a 32-bit adder is finite and is calculated by $\epsilon = 2^{32} \times 2^{32} = 18{,}446{,}744{,}073{,}709{,}551{,}616$. The generate, propagate, and kill signals that are derivatives of input data are responsible to determine the number of active levels of the adder, and the average-case delay of the shifter is determined by the total number of all possible input combinations ϵ. The set of critical paths $\mathcal{P} = \{\mathcal{P}_1, \mathcal{P}_2, \mathcal{P}_3\}$, with associated propagation delays $\Delta = \{\delta_1, \delta_2, \delta_3\}$ and the delay models in DGU, is designed with propagation delays $T = \{\tau_1, \tau_2, \tau_3\}$. The total number of paths $\gamma = 3$, three critical paths for capturing output from level 3, level 4 and level 5, respectively.

> The generate, propagate, and kill signals can be computed as soon as the input arrives at the adder, which in turn assist the adder in determining the number of active levels. Condition C1 established in Chap. 5 is satisfied for the adder as $\mathcal{G}_p(g, p, \mathit{k}) \mapsto \mathcal{P}_j$, i.e., active critical path of the adder can be determined by the generate, propagate, and kill signals, and hence a deterministic completion scheme can be developed for the binary adder.

Figure 6.8 illustrates the block diagram of a single-precision ABBA. ABBA consists of a single-precision Sklansky adder and a deterministic completion detection scheme, as discussed in the following sections.

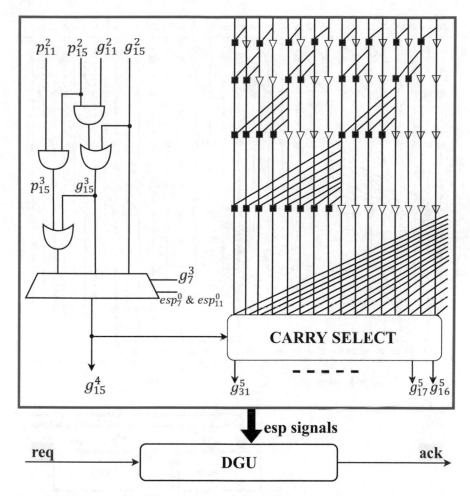

Fig. 6.8 Single-precision ABBA with deterministic completion detection scheme

6.3.1 Single-Precision Binary Adder

The binary adder used in this research is a hybrid of Sklansky and Carry Select Adders (CSAs). The traditional Sklansky adder is designed to generate all the carries simultaneously using generate and propagate signals, resulting in a fast adder architecture. The black and grey cells in the Sklansky adder are used recursively to generate 2-bit adders, 4-bit adders, 8-bit adders, 16-bit adders, and so on, by abutting two smaller adders each time, as shown in Fig. 6.9. The architecture is straightforward and consistent; it requires log_2n levels and needs fewer logic gates to implement an n-bit Sklansky adder, resulting in increased computation speed. White cells or buffers are utilised to improve the adder performance because it suffers from higher fan-outs [4].

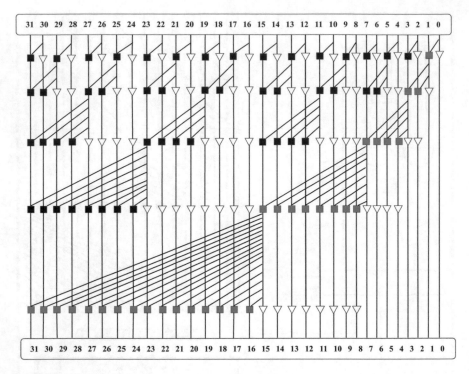

Fig. 6.9 32-bit Sklansky adder

Fig. 6.10 Carry select adder
(CSA)

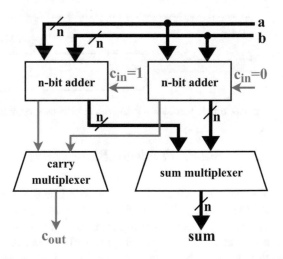

A CSA is designed using two non-overlapping sets of adders that generate two distinct adder outputs by considering both 0 and 1 input carry values [5]. The final values of sum and carry are selected later when the actual value of the input carry becomes known, as illustrated in Fig. 6.10. The CSA assists the deterministic completion detection circuit in determining which level generates the final output, while the remaining levels pass the output unchanged. The following section demonstrates the implementation of the hybrid binary adder in order to generate the *ack* signal after actual computation delay.

6.4   Deterministic Completion Detection Scheme for Single-Precision ABBA

Since ABBA does not have a clock signal, there is no worst-case delay constraint to assist the adder in commencing the next addition after the current addition is completed. Hence, to aid the ABBA in commencing its next addition operation, a completion detection scheme is required for determining when an ongoing addition process has completed. The objective of this study is to develop a deterministic completion detection scheme capable of detecting the completion of the addition operation after an actual computation delay corresponding to the actual input data and generating the *ack* signal accordingly.

As discussed in Chap. 5, the deterministic completion detection scheme is implemented using two blocks: DCDC and DGU. The following sections demonstrate the implementation of these blocks for an ABBA [3] using the proposed generic architecture.

6.4.1   DCDC for ABBA

As explained in Chap. 5, a DCDC consists of an IDS and an OSS. The DCDC for an ABBA is illustrated in Fig. 6.11. The speed of a specific adder architecture is determined by the propagation of the carry signal through the adder tree structure. A DCDC computes the carry within the tree structure by utilising the properties of the input data and/or their derivatives. This in turn would allow the ABBA to determine the actual computation delay required by the adder to compute the final sum, and DGU will generate the *ack* signal accordingly. The carry out from the ith bit when adding two numbers a and b is related to the level-0 generate, propagate, and kill signals of ith and $(i - 1)$th bits. The relation of these level-0 signals with the input data is illustrated in the following equation:

$$g_i^0 = a_i \cdot b_i,$$

$$p_i^0 = a_i \oplus b_i, \tag{6.9}$$

$$\ell_i^0 = (a_i + b_i)'.$$

Fig. 6.11 Deterministic
completion detection circuit
for single-precision ABBA

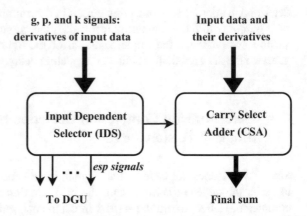

The propagate signal can be updated as $p_i^0 = a_i + b_i$, as $g_i^0 = a_i \cdot b_i$ covers the condition $a_i = b_i = 1$. The IDS is designed using these generate, propagate, and kill signals from distinct levels of the tree structured adder, and the carry out of the ith bit c_i can be calculated beforehand using the following equation:

$$c_i = \begin{cases} 0 & \text{when } p_i^0 \cdot \mathscr{k}_{i-1}^0 + \mathscr{k}_i^0 = 1, \\ 1 & \text{when } g_i^0 + p_i^0 \cdot g_{i-1}^0 = 1, \\ g_{i-2} & \text{when } p_i^0 \cdot p_{i-1}^0 = 1. \end{cases} \qquad (6.10)$$

The value of the carry out c_i is logic 0 when a kill term is computed as logic 1 at the ith or $(i-1)$th bit. The value of c_i logic 1 when the generate term is computed as logic 1 at the i bit or the generate bit at $(i-1)$th bit together with the propagate bit at ith bit is set to logic 1. The value of c_i may be logic 1 or logic 0, when both ith and $(i-1)$th bits can propagate the carry.

The group generate and propagate signals at various PPA levels can be determined by integrating the group generate and propagate signals from two smaller adjacent groups from the previous level. Equation 4.2 is modified to produce the group generate and propagate signals at a certain level ℓ of a PPA, as indicated by the following equation:

$$g_{k:l}^{\ell} = p_{k:i+1}^{\ell-1} \cdot g_{i:l}^{\ell-1} + g_{k:i+1}^{\ell-1},$$
$$p_{k:l}^{\ell} = p_{k:i+1}^{\ell-1} \cdot p_{i:l}^{\ell-1}. \tag{6.11}$$

Now, if the kill term at a level $\ell_i^0 = 1$, it implies that $g_i^0 = p_i^0 = 0$, and $p_i^0 = 0$ would result in $p_{i:l}^{\ell-1} = 0$, which leads to $p_{k:l}^{\ell} = 0$. For example, consider a 4-bit adder with the group propagation signal $p_{3:0}^{\ell} = p_{3:2}^{\ell-1} \cdot p_{1:0}^{\ell-1} = p_3^{\ell-1} \cdot p_2^{\ell-1} \cdot p_1^{\ell-1} \cdot p_0^{\ell-1}$. If $p_0^{\ell-1} = 0$, then $p_{3:0}^{\ell} = 0$; similarly, $g_i^0 = p_i^0 = 0$ would imply $g_{i:l}^{\ell-1} = 0$, which in turn gives $g_{k:l}^{\ell} = g_{k:i+1}^{\ell-1}$. Moreover, $\ell_i^0 = 1$ gives the condition $p_i^0 \cdot \ell_{i-1}^0 = 1$, i.e., bit i can propagate the carry, but the previous level cannot produce the carry. This condition provides the esp signal to enable shorter path if the addition operation requires less time than the worst-case delay, i.e., $esp_i = 1$ denotes that there is no carry at bit i. The following equation is used to compute the esp_i signal for implementing the carry chain in adder to determine its active levels:

$$esp_i = \ell_i^0 + p_i^0 \cdot \ell_{i-1}^0. \tag{6.12}$$

For a 32-bit Sklansky adder, bits 16 to 31 have the longest carry chain with the maximum delay of 5 levels. The generate signal for bits 16 to 31 at level 5 can be given as

$$g_{31:16}^5 = p_{31:16}^4 \cdot g_{15:0}^4 + g_{31:16}^4 \tag{6.13}$$

The $g_{15:0}^4$ drives the grey cells at level 5 and produces $g_{31:16}^5$. Now, the esp_{15} signal would determine the level to produce the final sum, as $esp_{15} = 0$ implies that all 5 levels are needed to produce the final sum, whereas $esp_{15} = 1$ implies that final sum can be captured at level 4. Equation 6.12 can be used to determine the esp_{15} signal as

$$esp_{15} = \ell_{15}^0 + p_{15}^0 \cdot \ell_{14}^0. \tag{6.14}$$

The signal $esp_{15} = 1$ indicates that the final sum is ready at level 4, and level 5 simply passes the output captured from level 4 without changing the final sum value. Therefore, the final sum can be captured at level 4, resulting in reduced computation delay by one level compared to the worst-case adder delay. The esp_{15} signal also assists the DGU to utilise the actual computation delay of four levels and generate the *ack* signal one-level delay early. The generate signal for bits 0 to 15 at level 4 can be given as

$$g_{15:0}^4 = p_{15:8}^3 \cdot g_{7:0}^3 + g_{15:8}^3. \tag{6.15}$$

The generate signal for bits 8 to 15 at level 3 can be further expanded in terms of level-2 generate and propagate signals, and hence $g_{15:0}^4$ can be written as

$$g_{15:0}^4 = p_{15:8}^3 \cdot g_{7:0}^3 + (p_{15:12}^2 \cdot g_{11:8}^2 + g_{15:12}^2). \tag{6.16}$$

Now, signals esp_{11} and esp_7 might further reduce the computation delay by one more level, as the final sum will be ready at level 3 if both esp_{11} and esp_7 are calculated as logic 1. The equations for esp_{11} and esp_7 are as follows:

$$\begin{aligned} esp_{11} &= \ell_{11}^0 + p_{11}^0 \cdot \ell_{10}^0, \\ esp_7 &= \ell_7^0 + p_7^0 \cdot \ell_6^0. \end{aligned} \tag{6.17}$$

A multiplexer is necessary to determine the level at which the final sum is ready and remaining levels have no impact on the final sum. The deterministic completion detection scheme is used with the carry select approach to enable shorter path if possible and avoid imposing a complete multiplexer delay on the adder tree structure. The following expression illustrates this condition to implement Pallavi's generic function as

$$g_{15}^4 = \begin{cases} g_{15}^2 & \text{when } esp_7 \cdot esp_{11} = 1, \\ g_{15}^3 & \text{when } g_7^3 = 0, \\ g_{15}^3 + p_{15}^3 & \text{when } g_7^3 = 1. \end{cases} \tag{6.18}$$

The implementation of $g_{15:0}^4$ multiplexer is illustrated in Fig. 6.12.

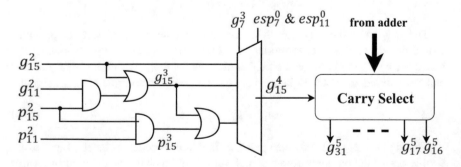

Fig. 6.12 Output selection stage for adder

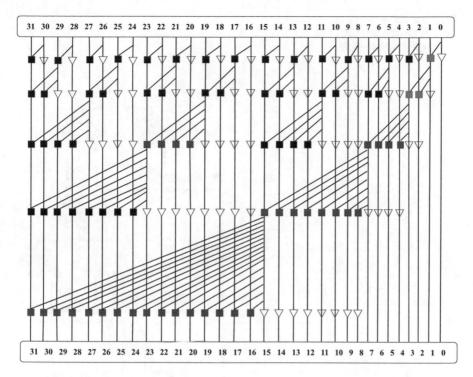

Fig. 6.13 32-bit hybrid binary adder [3]

The ABBA is implemented using a hybrid adder architecture that combines carry select and deterministic completion detection structures and substitutes the grey cells in the Sklansky adder, as illustrated in Fig. 6.13 highlighted in blue.

6.4.2   DGU for ABBA

DGU is responsible for asserting the *ack* signal after propagation delay associated with the active critical path of the adder. As mentioned in Chap. 5, DGU contains multiple delay models that are designed to replicate the propagation delays of the various signal paths present in the adder. The most common way of implementing a delay model is by utilizing a chain of inverters, and however the optimal solution for creating matching delay models for DGU is to use the same logic elements that implement the signal paths of the adder. Since the adders (Sklansky and CSA) are built using 2-level grey cells, black cells, and multiplexers, the delay models in DGU are also implemented using basic gates designed to function like adder modules. The *esp* signals are combined strategically to generate the distinct computation delays of the adder corresponding to its active signal path. The *esp* signal indicates whether

Fig. 6.14 Delay models in delay generating unit (DGU) for single-precision ABBA

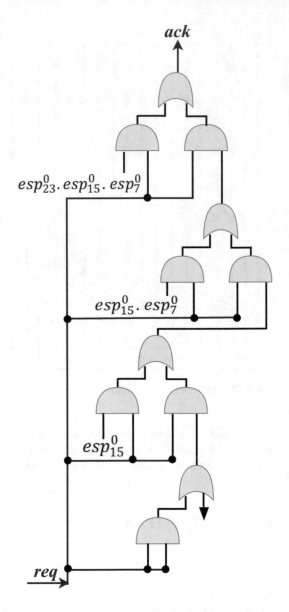

the computation will need the worst-case delay, or it will finish the computation before it. Three conditions are defined using esp signals, and if any of the conditions is set to logic 1, a shorter path is activated, as shown in Fig. 6.14.

A 32-bit Sklansky adder is implemented using five levels, and if the carry out from bit 15 is logic 0, i.e., $c_{15} = 0$, it implies that $g^5_{16:31} = g^4_{16:31}$. In other words, the sum produced at level 4 will remain unchanged at level 5 for $esp_{15} = 1$, and the final sum can be captured at level 4. Similarly, the final sum can be captured at level 3 if

$esp_7 = 1$ and $esp_{23} = 1$ together with $esp_{15} = 1$, as the output produced at level 3 will remain unchanged at level 4 and level 5. The delay required to implement the logic to eliminate level 3 and earlier levels is comparable to or greater than the delay required to implement the adder itself, and hence implementing the logic beyond eliminating level 4 becomes impractical and is therefore not developed.

6.5   Chapter Summary

This chapter presents the design examples for implementing asynchronous bundled data barrel shifter and binary adder using the generic approach proposed in Chap. 5. The architectures of Deterministic Completion Detection Circuit (DCDC) and Delay Generating Unit (DGU) were developed for the shifter and adder such that the completion of the computation (i.e., shifting and addition, respectively) can be detected after actual computation delay. To emulate the actual circuit delay used in ABBS and ABBA stages, the delay models were developed using basic gates configured to operate as shifter and adder stages, respectively. The delay models were designed with more inputs than the actual architecture because the DGU must be slower than the actual circuit to enable error-free operation. The simulation result of these two architecture is provided in the next chapter.

References

1. Srivastava, P. and Chung, E. (2022), 'An asynchronous bundled-data barrel shifter design that incorporates a deterministic completion detection technique', IEEE Transactions on Circuits and Systems II: Express Briefs 69(3), 1667–1671.
2. Srivastava, P., Chung, E. and Ozana, S. (2020), 'Asynchronous floating-point adders and communication protocols: A survey', Electronics 9(10), 1687.
3. Lai, K. K. (2016), Novel Asynchronous Completion Detection For Arithmetic Datapaths, Taylor's University, Malaysia.
4. Jain, A., Bansal, S., Akhter, S. and Khan, S. (2020), Vedic-based squaring circuit using parallel prefix adders, in '2020 7th International Conference on Signal Processing and Integrated Networks (SPIN)', IEEE, pp. 970–974.
5. Parhami, B. (2010), Computer arithmetic: Algorithms and hardware designs, Vol. 20, Oxford University Press.

Chapter 7
Simulations

7.1 Evaluation of a Single-Precision Barrel Shifter

The experiments for this study were carried out using Xilinx Vivado® Design Suite in a simulation environment. Both CBS and ABBS designs described in this study were implemented using Verilog HDL, with gate-level delays included to simulate the behavior of real devices. The maximum propagation delay of the 1x drive strength for AND, OR, and NOT gates utilised in the barrel shifter design is summarised in Table 7.1, which was taken from Samsung Electronics' 0.35 μm 3.3V CMOS standard cell library [1], in order to emulate a netlist in this library. The propagation delay of a logic gate is the delay required for the data to travel from its input to output, and it varies depending on which input pin is used to determine its value. This study took a pessimistic approach by using the maximum propagation delay for a logic gate among its input pins, which emulates the critical condition for asynchronous design style. Figure 7.1 illustrates the ABBS architecture designed using Pallavi's Generic Function G_p.

A 2×1 multiplexer is the key module of a CBS, which requires one inverter (m_1), two AND gates (m_2 & m_3), and one OR gate (m_4), as illustrated in Fig. 7.2. The netlist of CBS is designed with the gate delay values taken from Table 7.1 for the 0.35 μm library, as shown in Fig. 7.3. The worst-case delay of the 2×1 requires three gate delays which include gates m_1, m_2, and m_4.

The clock signal clk in the test bench is inverted after the minimum delay required to complete the process, in order to compute the optimised worst-case delay. Given that the gate delays present in the critical path are taken from Table 7.1, the estimated delay is between 1000 ps and 1400 ps, after which the clk should be inverted. Accordingly, different values of clk were tested within this range, and it was experimentally observed that if the clk value is inverted after 1180 ps, the CBS is able to complete the shifting process but fails if this value is 1179 ps; even a 1 ps

© The Author(s), under exclusive license to Springer Nature Switzerland AG 2022
P. Srivastava, *Completion Detection in Asynchronous Circuits*,
https://doi.org/10.1007/978-3-031-18397-3_7

Table 7.1 Gate delay for 1x drive [1]

AND gate				
Propagation delay (ps)	2-input AND	3-input AND	4-input AND	5-input AND
$\delta_{plh\text{-}max}$	185	215	242	248
$\delta_{phl\text{-}max}$	197	232	268	309
OR gate				
Propagation delay (ps)	2-input OR	3-input OR	4-input OR	5-input OR
$\delta_{plh\text{-}max}$	162	189	210	200
$\delta_{phl\text{-}max}$	257	338	429	366
Inverter				
Propagation delay: $\delta_{plh} = 110$ ps and $\delta_{phl} = 90$ ps				

Fig. 7.1 ABBS using Pallavi's Generic Function $\mathcal{G}_p$

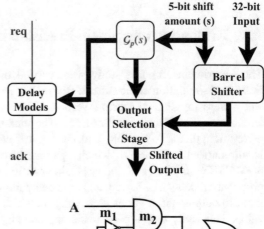

Fig. 7.2 2×1 Multiplexer

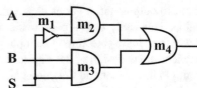

difference would cause output to be captured prematurely, and the worst-case delay of the CBS is calculated in Eq. 7.1 as

$$\delta_{CBS} = 1180 \times 2 = 2360 \text{ ps} = 2.36 \text{ ns}. \tag{7.1}$$

The output of a CBS designed with a clock pulse must be captured after a delay of at least 2.36 ns. This minimum value of computation delay is fixed for the clocked CBS, irrespective of active shifter stages associated with the input, as illustrated in Fig. 7.5a.

Likewise, the netlist for an ABBS is also designed with gate delay values taken from Table 7.1, as shown in Fig. 7.4. The delay models in DGU are developed using basic gates to replicate the multiplexer architecture used in CBS stages and OSS, with more input pins than the actual shifter. This is done to ensure that the

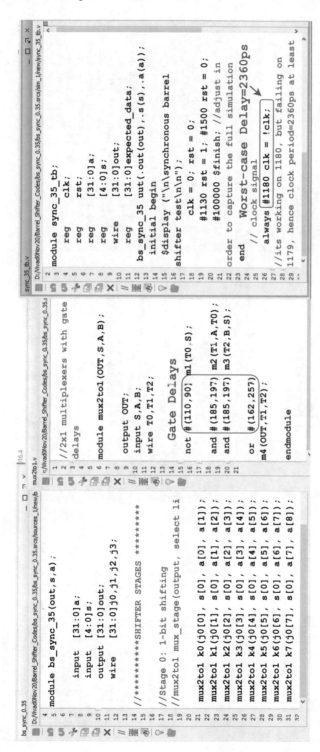

Fig. 7.3 CBS simulation code window

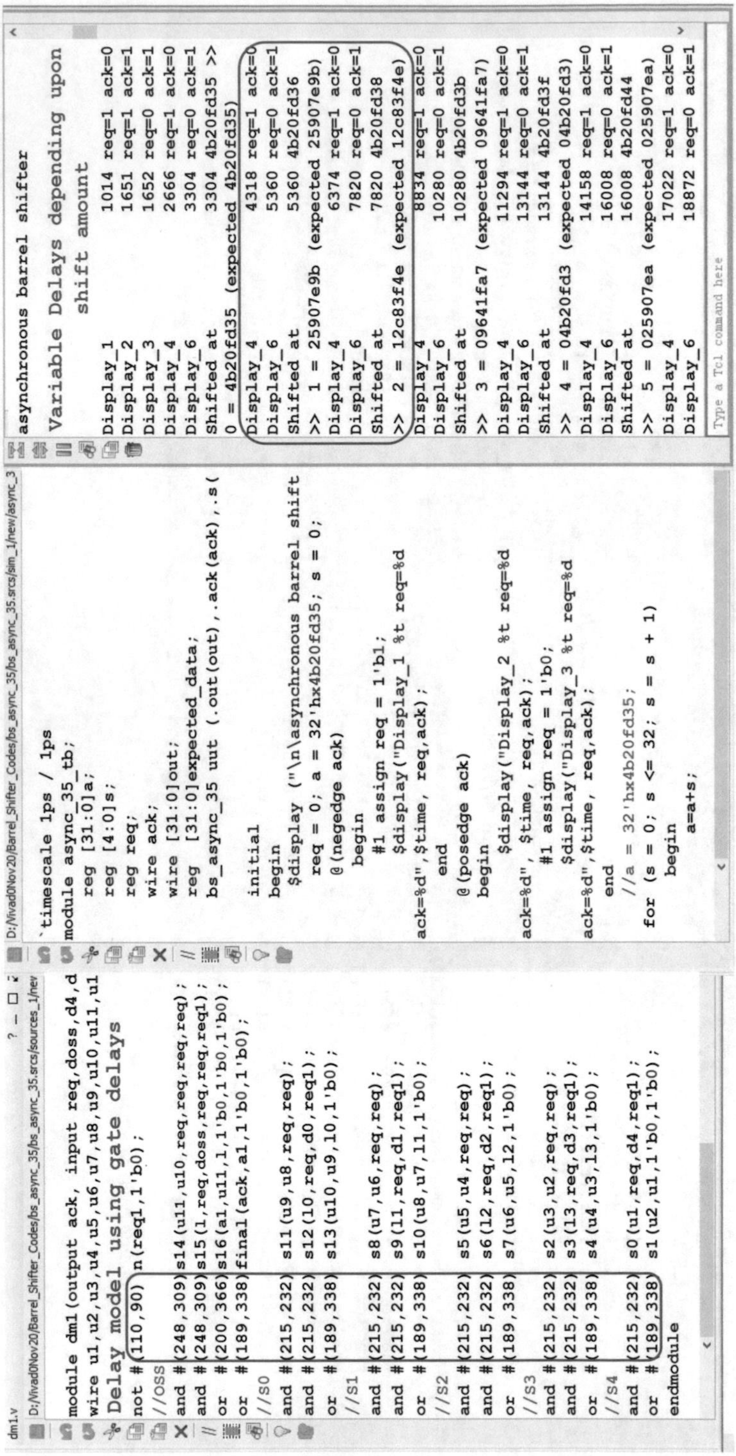

dm1.v

D:/Nvad0Nov20/Barrel_Shifter_Codes/bs_async_35.srcs/sources_1/new

```
module dm1(output ack, input req,doss,d4,d
wire u1,u2,u3,u4,u5,u6,u7,u8,u9,u10,u11,u1
Delay model using gate delays
not #(110,90) n(req1,1'b0);
//OSS
and #(248,309)s14(u11,u10,req,req,req,req);
and #(248,309)s15(1,req,doss,req,req,req1);
or #(200,366)s16(a1,u11,1,1'b0,1'b0,1'b0);
or #(189,338)final(ack,a1,1'b0,1'b0);
//S0
and #(215,232) s11(u9,u8,req,req);
and #(215,232) s12(10,req,d0,req1);
or  #(189,338) s13(u10,u9,10,1'b0);
//S1
and #(215,232) s8(u7,u6,req,req);
and #(215,232) s9(11,req,d1,req1);
or  #(189,338) s10(u8,u7,11,1'b0);
//S2
and #(215,232) s5(u5,u4,req,req);
and #(215,232) s6(12,req,d2,req1);
or  #(189,338) s7(u6,u5,12,1'b0);
//S3
and #(215,232) s2(u3,u2,req,req);
and #(215,232) s3(13,req,d3,req1);
or  #(189,338) s4(u4,u3,13,1'b0);
//S4
and #(215,232) s0(u1,req,d4,req1);
or  #(189,338) s1(u2,u1,1'b0,1'b0);
endmodule
```

D:/Nvad0Nov20/Barrel_Shifter_Codes/bs_async_35.srcs/sim_1/new/async_3

```
`timescale 1ps / 1ps
module async_35_tb;
reg [31:0]a;
reg [4:0]s;
reg req;
wire ack;
wire [31:0]out;
reg [31:0]expected_data;
bs_async_35 uut (.out(out),.ack(ack),.s(

initial
begin
$display ("\nasynchronous barrel shift
req = 0; a = 32'hx4b20fd35; s = 0;
@(negedge ack)
   begin
     #1 assign req = 1'b1;
     $display("Display_1 %t req=%d
ack=%d",$time, req,ack);
   end
@(posedge ack)
   begin
     $display("Display_2 %t req=%d
ack=%d", $time, req,ack);
     #1 assign req = 1'b0;
     $display("Display_3 %t req=%d
ack=%d",$time, req,ack);
   end
//a = 32'hx4b20fd35;
for (s = 0; s <= 32; s = s + 1)
   begin
     a=a+s;
```

```
asynchronous barrel shifter

Variable Delays depending upon
           shift amount

Display_1          1014 req=1 ack=0
Display_2          1651 req=1 ack=1
Display_3          1652 req=0 ack=1
Display_4          2666 req=1 ack=0
Display_6          3304 req=1 ack=1
Shifted at         3304 4b20fd35 >>
0 = 4b20fd35  (expected 4b20fd35)

Display_4          4318 req=1 ack=0
Display_6          5360 req=1 ack=1
Shifted at         5360 4b20fd36
>> 1 = 25907e9b (expected 25907e9b)
Display_4          6374 req=1 ack=0
Display_6          7820 req=1 ack=1
Shifted at         7820 4b20fd38
>> 2 = 12c83f4e (expected 12c83f4e)
Display_4          8834 req=1 ack=0
Display_6         10280 req=0 ack=1
Shifted at        10280 4b20fd3b
>> 3 = 09641fa7 (expected 09641fa7)
Display_4         11294 req=1 ack=0
Display_6         13144 req=0 ack=1
Shifted at        13144 4b20fd3f
>> 4 = 04b20fd3 (expected 04b20fd3)
Display_4         14158 req=1 ack=0
Display_6         16008 req=0 ack=1
Shifted at        16008 4b20fd44
>> 5 = 025907ea (expected 025907ea)
Display_4         17022 req=1 ack=0
Display_6         18872 req=0 ack=1

Type a Tcl command here
```

Fig. 7.4 ABBS simulation code window

propagation delay of the DGU is always greater than the actual computation delays of CBS and OSS. As mentioned previously, using logic gates with weaker drive strength than the actual gates in the shifter is an alternative technique for creating slower delay models.

Figure 7.5b illustrates the shift-dependent computation time of ABBS. The propagation delay of ABBS δ_{ABBS} corresponding to the shift amount can be calculated using the Tcl Console window shown in Fig. 7.4 or from the waveform shown in Fig. 7.5b by finding the difference between the time when the ack signal changes from low to high and the time when the req signal changes from low to high for the given shift amount. Hence, the propagation delays for shift amounts $s = 0, s = 1, s = 2, 3 s = 4–7, 8–15$, and $s = 16–32$ for the 0.35 μm library are computed as $\delta_1 = 638$ ps, $\delta_2 = 1042$ ps, $\delta_3 = 1446$ ps, $\delta_4 = 1850$ ps, $\delta_5 = 2254$ ps, and $\delta_6 = 2658$ ps, respectively. The delay of ABBS can be calculated from Eq. 6.8 as

$$\delta_{ABBS} = \delta_1 \times \left(\frac{1}{32}\right) + \delta_2 \times \left(\frac{1}{32}\right) + \delta_3 \times \left(\frac{1}{16}\right) + \delta_4 \times \left(\frac{1}{8}\right) + \delta_5 \times \left(\frac{1}{4}\right) + \delta_6 \times \left(\frac{1}{2}\right)$$

$$= 638 \times \left(\frac{1}{32}\right) + 1042 \times \left(\frac{1}{32}\right) + 1446 \times \left(\frac{1}{16}\right) + 1850 \times \left(\frac{1}{8}\right) + 2254 \times \left(\frac{1}{4}\right)$$

$$+ 2658 \times \left(\frac{1}{2}\right) = 2266.63 \text{ ps} \approx 2.27 \text{ ns.}$$

$$(7.2)$$

Figure 7.5 illustrates the simulation waveform for CBS and ABBS designed using 0.35 μm library. The CBS configured with a clock signal initiates the shifting process on the rising edge of clock signal clk, and the shifted data will be available prior to the next rising edge of clk. The time period of the clk signal is fixed at 2.36 ns irrespective to the shift amount s, and a new shifting process can be initiated at every rising edge of clk, as shown in Fig. 7.5a.

On the contrary, ABBS initiates the shifting process on the rising edge of request signal req, and the shifted data will be available prior to the rising edge of acknowledge signal ack. The ack signal is asserted after the actual computation delay required to shift the data for a given shift amount. The rising edge of ack signal indicates that the final shifted output is ready, and a new shifting process can be initiated at the next rising edge of req, as shown in Fig. 7.5b. For instance, data 32'h × 4b20fd35 needs to be shifted by 1-bit on the rising edge of req signal at 4318 ps, and the ack signal is asserted at 5360 ps, indicating that the final shifted data (32'h × 25907e9b) will be available prior to rising edge of the ack signal. Similarly, data 32'h × 4b20fd38 needs to be shifted by 2-bit on the rising edge of req signal at 6374 ps, and the ack signal is asserted at 7820 ps, indicating that the final shifted data (32'h × 12c83f4e) will be available prior to rising edge of the ack signal. The test benches for CBS and ABBS for 0.35 μm library are provided in Appendix B, in Figs. B.1 and B.2, respectively.

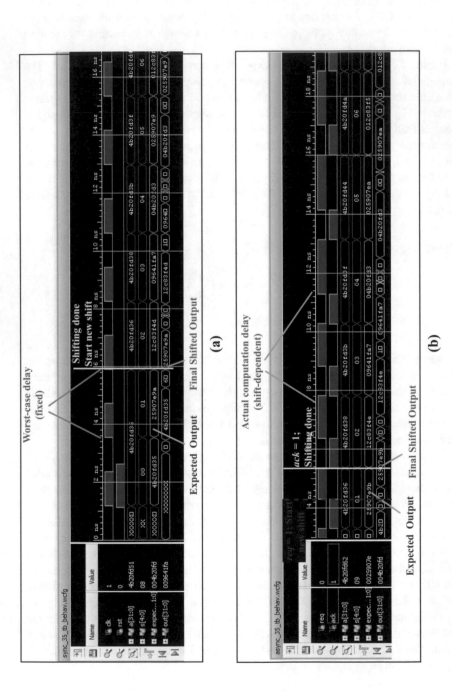

Fig. 7.5 Simulation waveform for (**a**) CBS (**b**) ABBS

7.1.1 An Illustrative Example

Equation 7.2 provides the theoretical computation delay for a 32-bit ABBS for the various shift values, with the assumption that there is an equal probability of occurrence for all shift values. It appears from Eqs. 7.1 and 7.2 that the average-case delay of the ABBS is equivalent to the worst-case delay of CBS. However, in real-world applications, the distribution of shift values is not uniform. For instance, shifting is a key operation in floating-point adders (FPAs), which is commonly utilised in real-world applications [2]. The adder requires a right shifter for matching the exponents before performing the mantissa addition and a left shifter for normalising the result after the addition. The basic architecture of an FPA is discussed in Appendix A. Oberman [3] examined the data for ten applications from the SPEC Benchmark Release [4], each of which executed about three billion instructions on a DEC Alpha 3000/500 workstation for standard input datasets. This evaluation shows that the exponent matching process requires to right shift the data by $s \leq 1$ for 43% of operations, whereas the normalisation process requires to left shift the data by $s \leq 2$ for 52.5% of the operations, as illustrated in Figs. 7.6 and 7.7, respectively.

Comparing these results with another study of floating-point addition operation on a different architecture and with different applications yields fairly similar results. While evaluating the aligning and normalising shift distances for a certain research on an IBM 704 over 60 years ago, six issues were discovered. In 45% of the operands, a 0 or 1 bit right shift was necessary, whereas in 55% of cases, more than a 1-bit right shift was necessary. The consistency of the results over such a long period of time demonstrates a fundamental distribution of floating-point addition operands in scientific applications [3]. Figure 7.6 provides the distribution

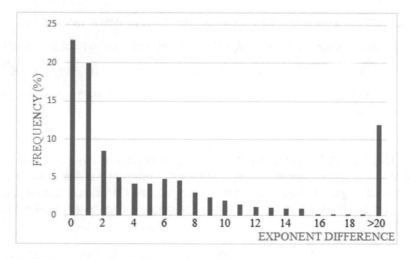

Fig. 7.6 Histogram of exponent difference [3]

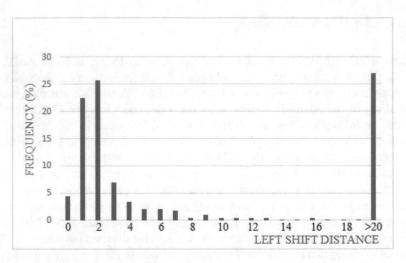

Fig. 7.7 Histogram of normalising shift distance [3]

of the shift amount required by FPA for exponent matching process. Therefore, the computation delay of the shifter for performing right shift for exponent matching in FPA can be calculated by Eq. 2.2 as

$$\delta_{em} = \sum_{i=1}^{\gamma_c} (\delta_i . prob(e_i)) . \tag{7.3}$$

The values of probability of the input data from Fig. 7.6 and its corresponding delay obtained from the simulation are put in Eq. 7.3 to calculate the value of δ_{em}

$$\delta_{em} = 638 \times 0.23 + 1042 \times 0.20 + 1446 \times 0.085 + 1446 \times 0.05 + 1850 \times 0.042$$

$$+ 1850 \times 0.042 + 1850 \times 0.048 + 1850 \times 0.046 + 2254 \times 0.03 + 2254 \times 0.024$$

$$+ 2254 \times 0.02 + 2254 \times 0.015 + 2254 \times 0.012 + 2254 \times 0.01 + 2254 \times 0.009$$

$$+ 2254 \times 0.009 + 2658 \times 0.002 + 2658 \times 0.002 + 2658 \times 0.002 + 2658 \times 0.002$$

$$+ \frac{2658}{12} \times 0.12 = \mathbf{1194.734\,ps} \approx \mathbf{1.19\,ns}.$$

$$\tag{7.4}$$

A 32-bit ABBS for performing left shift can be developed using the same logic as the right shifter proposed in Chap. 6. Figure 7.7 provides the distribution of the shift amount required by FPA for normalisation process. Therefore, the computation delay of the shifter for performing left shift for normalisation process in FPA can be calculated by Eq. 2.2 as

$$\delta_{norm} = \sum_{i=1}^{\gamma_c} (\delta_i . prob(e_i)) . \tag{7.5}$$

The values of probability of the input data from Fig. 7.7 and its corresponding delay obtained from the simulation are put in Eq. 7.5 to calculate the value of δ_{norm}

$$\delta_{norm} = 638 \times 0.044 + 1042 \times 0.224 + 1446 \times 0.257 + 1446 \times 0.07 + 1850 \times 0.034$$

$$+ 1850 \times 0.021 + 1850 \times 0.021 + 1850 \times 0.018 + 2254 \times 0.005 + 2254 \times 0.01$$

$$+ 2254 \times 0.004 + 2254 \times 0.004 + 2254 \times 0.004 + 2254 \times 0.005 + 2254 \times 0.001$$

$$+ 2254 \times 0.001 + 2658 \times 0.004 + 2658 \times 0.001 + 2658 \times 0.001 + 2658 \times 0.001$$

$$+ \frac{2658}{12} \times 0.27 = \mathbf{1063.269\,ps} \approx \mathbf{1.06\,ns}.$$

(7.6)

As demonstrated by Eqs. 7.1 and 7.4, when a 32-bit right-shift barrel shifter is designed using the proposed generic architecture, the computation delay for exponent matching operation in FPA is reduced from 2.36 ns to 1.19 ns, resulting in a **49.57%** performance improvement. Similarly, comparing Eqs. 7.1 and 7.6 demonstrates that the computation delay of a 32-bit left-shift barrel shifter designed using the proposed generic architecture is reduced from 2.36 ns to 1.06 ns for normalisation process in FPA, which provides a performance improvement of **55.08%**. This is a major performance boost in the shifter design, despite the propagation delays of additional logic for deterministic completion detection of computation.

The simulations for the CBS and ABBS architectures are also performed using 90 nm library [5] to demonstrate that the proposed design is not library-specific in terms of performance. Table 7.2 details the CBS delay (δ_{CBS}), ABBS delay (δ_{ABBS}), and delay of ABBS in FPA application for exponent matching (δ_{em}) and normalization (δ_{norm}) for 0.35 μm [1] and 90 nm [5] libraries.

The area, power, delay, and power delay product (PDP) of the synchronous and asynchronous barrel shifters are calculated as described in [6] and summarised in Table 7.3 for 90 nm standard cell library [7].

As previously stated, large shifts are uncommon in real-time applications; thus, the average-case delay of an asynchronous barrel shifter provides better performance than the worst-case delay of its synchronous counterpart. While the proposed ABBS design requires a larger implementation area, it outperforms its synchronous counterpart in terms of speed and power consumption when used in FPA, as illustrated in Table 7.3.

Table 7.2 Delay of barrel shifter for synchronous and asynchronous design style

Library	δ_{CBS} (ns)	δ_{ABBS} (ns)	δ_{norm} (ns)	δ_{em} (ns)
90 nm	1.86	2.19	1.02	1.17
0.35 μm	2.36	2.27	1.06	1.19

Table 7.3 Performance comparison of 32-bit CBS and proposed ABBS

Shifter design	Delay (ns)	Area (μm^2)	Power (μW)	PDP ($\mu W.ns$)
Conventional (CBS)	1.86	4571.14	35.52	65.93
Proposed (ABBS)	2.19	7958.02	51.29	112.34
ABBS for FPA (Normalization)	1.02	7958.02	25.54	26.05
ABBS for FPA (Exponent matching)	1.17	7958.02	30.99	36.25

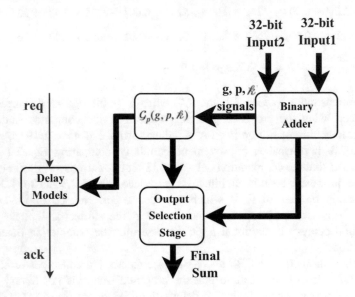

Fig. 7.8 ABBA using Pallavi's Generic Function $\mathcal{G}_p$

7.2 Evaluation of a Single-Precision Binary Adder

Figure 7.8 illustrates the ABBA architecture designed using Pallavi's Generic Function $\mathcal{G}_p$. The adder designed by Lai [8] demonstrates the perfect customisation of the proposed generic architecture. Synopsys Verilog Compiler and Simulator is used for simulating the ABBA RTL, and netlist is generated using Synopsys Design Compiler. The static timing analysis and formal verification of the ABBS design is done using Synopsys PrimeTime and Synopsys Formality, respectively. The ABBA is implemented in Verilog HDL with a 100 ps timescale and by instantiating Synopsys GTECH cells with the standard delay. The classic binary adder and ABBA, both designs are simulated with a total of 2,147,483,648 vectors, and RTL simulation is used to benchmark the average-case performance. Figure 7.9 demonstrates how the test bench of ABBA is used to generate different combinations of input operands.

```
`timescale 100ps / 1ps

module adder_tb_top ();

reg [31:0] a;
reg [31:0] b;
reg        req;
wire       ack;
wire [31:0] s;
reg [31:0] check_sum;

adder uadder (.op1(a), .op2(b), .sum(s), .req(req), .ack(ack));

initial
begin
  $display ("\n\n Sklansky adder version 1.0 \n\n");
  assign req = 0;
  for (a = 0; a <= 1073741824; a = a+1)begin
    for (b = 0; b<= 1; b = b+1)begin
      assign check_sum = a+b;
      @(negedge ack) begin
        #1 assign req = 1`b1;
    end
      @(posedge ack) begin
        #1 assign req = 1`b0;
        if (s != check_sum)
        $display ("%t %d     +%d   = %d  fail", $time,a,b,s);
    end
    end
  end
  $finish;
end
endmodule
```

Input operand a = 0 to 1000 0000 0000 0000 0000 0000 0000 0000 in binary; Input operand b toggles from 0 to 1 whenever operand a is incremented

Checker

Fig. 7.9 ABBA code window for test bench

The carry selection in the proposed ABBA is done using different groups of carry select multiplexer, as illustrated in Fig. 7.10. A GTECH_MUX4 is used to design the carry select multiplexer, making one of the input data redundant to implement a 3×1 multiplexer. Since asynchronous design is not supported by the Design Compiler, ABBA is synthesised in a synchronous environment by replacing a single clock with *req* and *ack* signals to signify the start and end of the computation process, respectively. The GTECH cells are instantiated to prevent the Design Compiler from replacing the adder with Design Ware and to retain the adder's topology.

Carry selections for group 1 are produced using early detection of bits [3:2], and the carry select multiplexer is positioned at level 3 of ABBA. Likewise, the carry select multiplexer for group 2 is positioned at level 4. As discussed in Sect. 6.4.1, if the value of carry generated from bit 7, level 3 of a Sklansky adder is logic 0,

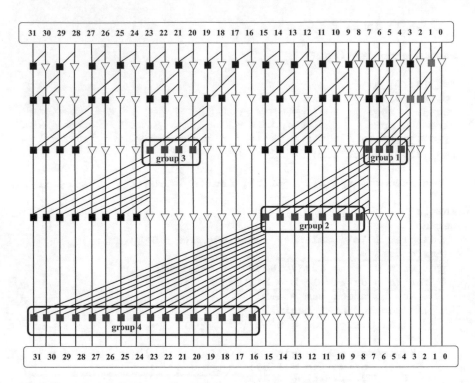

Fig. 7.10 32-bit hybrid binary adder [8]

i.e., $g_7^3 = 0$, then $g_{15:8}^4 = g_{15:8}^3$. If this condition is combined with esp signals such that esp_7 and esp_{11} both have logic 1 value, then the final sum will be available two-level delay earlier than the worst-case delay. However, if the value of carry generated from bit-7, level-3 of a Sklansky adder is logic 0, i.e., $g_7^3 = 1$, then an additional delay of an OR gate is introduced in the overall adder delay, as illustrated in Eq. 7.7.

$$g_{15:8}^4 = \begin{cases} g_{15:8}^2 & \text{when } esp_7 \cdot esp_{11} = 1, \\ g_{15:8}^3 & \text{when } g_7^3 = 0, \\ g_{15:8}^3 + p_{15:8}^3 & \text{when } g_7^3 = 1. \end{cases} \tag{7.7}$$

Equation 7.7 provides three different values of the generate signal $g_{15:8}^4$, which depends upon the value of input operands. For instance, the group generate signal $g_{15:8}^4$ of ABBA can be captured two-level delay earlier than the worst-case delay, if the input operand provides the pattern for propagate signal as $p_{31}p_{30}p_{29}p_* \ldots \not\!k_{11}p_{10}p_9p_8\not\!k_7p_6p_5p_* \ldots p_0$. This condition yields the optimal response from ABBA, as determining the final sum prior to two levels would require more literals to implement the deterministic completion detection network, causing

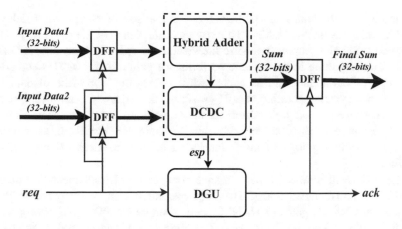

Fig. 7.11 ABBA integration

the DCDC to run slower than the worst-case delay, resulting the adder to always operate at the worst-case delay.

Similarly, if the pattern $p_{31} p_{30} p_{29} p_* \cdots p_{11} p_{10} p_9 p_8 \bar{k}_7 p_6 p_5 p_* \cdots p_0$ is generated from the input operands, then the group generate signal $g_{15:8}^4$ of ABBA can be captured one-level delay earlier than the worst-case delay. However, pattern $p_{31} p_{30} p_{29} p_* \cdots p_{11} p_{10} p_9 p_8 g_7 p_6 p_5 p_* \cdots p_0$ introduces an additional delay of an OR gate in the carry select multiplexer due to the condition $g_{15:8}^4 = g_{15:8}^3 + p_{15:8}^3$ when $g_7^3 = 1$. The optimisation of carry select multiplexers of group 3 and group 4 is done in a similar manner at level 3 and level 5, respectively. D flip-flops are used to ensure that input data arrives at the same time with the req signal, and ack signal is generated after the output is ready. This configuration will assist the DGU in preventing the premature generation of the ack signal, hence avoiding the hazards. The implementation of ABBA using D flip-flops and handshaking signals is depicted in Fig. 7.11.

7.2.1 Results of ABBA

As mentioned earlier, approximately 2×10^9 vectors are used to validate the ABBA performance. The ABBA is able to generate the ack signal after the sum is valid and stable, indicating that the ABBA can perform the addition operation accurately. Moreover, since the ABBA introduces an additional delay to the critical path, both the binary adder with clock and ABBA are analysed, which provides that only one inversion gate delay is added extra to the critical path of ABBA. ABBA and the conventional binary adders operate at maximum frequencies of 714 MHz and 741 MHz, respectively, when the clock period is restricted to 1.2 ns.

Figure 7.12 illustrates the performance comparison of computation delay of ABBA and conventional adder. The computation delay of the Sklansky adder designed using the proposed generic architecture is reduced from 3865.47 ms to 3449.13 ms, resulting in an absolute performance gain of 416 ms. This equates to a savings of about 297 million cycles when computing 2 gigabits of data in a synchronous system operating at a 714 MHz clock frequency, with a time period of 1.4 ns. More precisely, the deterministic completion detection scheme for the ABBA designed using Sklansky adder has an additional advantage, as it utilises the high fan-out topology of the Sklansky adder to implement Pallavi's generic function $\mathcal{G}_p$.

Figure 7.13 illustrates the performance comparison of implementation area of ABBA and conventional adder. The implementation area is computed in terms of two-input NAND gates or NAND2, with a single NAND2 gate having a size of 15.27 units. Silterra 0.13 um Sage-X process technology is used to synthesise the RTL of ABBA to a gate-level netlist in order to compare the power and gate size. The Sklansky adder designed using the proposed generic architecture has a smaller footprint than its conventional version by roughly 20 NAND2 gates. When compared to the size of the generic 32 bit SK adder, which is around 490 NAND2 gates, this difference of 20 NAND2 gates is negligible.

Figure 7.14 illustrates the performance comparison of power consumed by ABBA and conventional adder. Silterra 0.13um Sage-X process technology is used to synthesise the RTL of ABBA to a gate-level netlist in order to compare the power and gate size. The difference between the power consumed by an ABBA and a

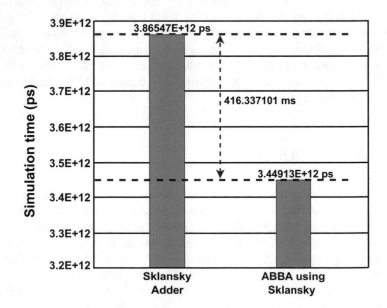

Fig. 7.12 Performance comparison computation delay

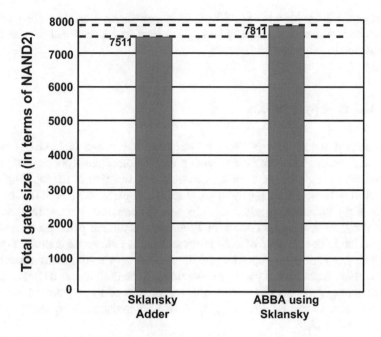

Fig. 7.13 Performance comparison implementation area

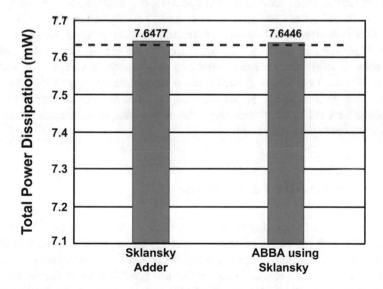

Fig. 7.14 Performance comparison power consumption

conventional binary adder is insignificant. It is worth noting that the ABBA with the SKlansky adder consumes slightly less power than its conventional version due to the reduction in cell power.

7.3 Other Applications

The proposed model in this book is validated using two key ALU modules: shifter and adder. The shifter and adder architectures are systolic arrays, i.e., they independently evaluate a partial outcome as a function of the input operands received from their upstream neighbors, store it internally, and pass it downstream. Systolic array architectures are frequently used in computer architecture; they are implemented in stages and consist of finite unique critical paths associated with the input data. The number of unique critical paths connecting inputs to outputs is limited, even for a large number of input data; as a result, delay models that replicate these finite critical paths are also finite. Such designs can be built utilising the proposed generic architecture, as the completion of a computation process in such architectures can be detected after actual computation delay using Pallavi's generic function G_p.

Furthermore, adder and shifter architectures serve as the basic building blocks for a variety of other computational circuits. Multiplier can be designed using both adder and shifter. Increment/decrement operators also require adder. Floating-point adder needs both adder and shifter to complete its operation. FPA is a critical block of a processor as FP operations are required for accurate calculations [2]. Hence implementing adder and shifter will provide the fundamental module of the processor, using which several other computation blocks can be designed. The absence of global clock assists such circuits to operate on an average-case delay, without facing clock-related issues and can optimize the power consumption, as the modules are active only when they process the data, else the circuit is idle and minimal power will be consumed.

7.4 Suggestions for Future Work

In this research, multiple delay models are used to implement the Delay Generating Unit (DGU), with the condition that the propagation delay of the delay models in DGU must be slower than the propagation delay associated with the active critical path of the computation block. To ensure that this condition is met, the delay models developed in this book utilise more input pins than the computation block itself, as mentioned in Chap. 7. It is worth noting that further research is still needed to fully understand how to best match the propagation delay of the delay model to the actual circuit it is attempting to replicate.

Chapter 1 highlighted that electronics industries prefer and commercialise synchronous circuits because the synchronous design style is supported by a wide range of Electronic Design Automation (EDA) tools. While EDA tools for asynchronous design are being developed, customising synchronous EDA tools to simulate and synthesise asynchronous circuits is still a common practice. Hence, future research in the area of developing EDA tools for asynchronous paradigms is required.

Until now, this research has been focused on developing a generic architecture for a deterministic completion detection scheme with specific conditions. ABBS and ABBA are the outcome of this effort. Several other asynchronous designs can be implemented using this generic architecture, provided certain conditions as defined in Chap. 5 are met. Furthermore, regardless of whether simulation results justify the proposed scheme's correctness, the ultimate proof of correctness remains physical, i.e., the design of a real-world system. While the book referred to some real-world design experience with ABBS and ABBA, more are planned.

7.5 Chapter Summary

This chapter presents the experimental results for the designs proposed in Chap. 6. The Conventional Barrel Shifter (CBS) and the proposed Asynchronous Bundled Data Barrel Shifter (ABBS) are both simulated using the Xilinx Vivado® Design Suite, and their results are compared. Furthermore, the performance of the ABBS is evaluated by considering the probability of the shift amount for real-time applications, and its average-case delay is calculated. When compared to the performance of CBS, the ABBS exhibits a significant improvement in real-time applications. Additionally, an ABBA design is utilised to validate the proposed generic architecture for deterministic completion detection, which outperforms the conventional binary adder. The chapter is concluded with the suggestions for future work.

References

1. Samsung, E. C. L. (2000), 0.35 μm3.3 v CMOS standard cell library for pure logic.', MDL Products Databook.
2. Srivastava, P., Chung, E. and Ozana, S. (2020), 'Asynchronous floating-point adders and communication protocols: A survey', Electronics 9(10), 1687.
3. Oberman, S. F. (1996), Design Issues in High Performance Floating Point Arithmetic Units, Citeseer.
4. SPEC Benchmark Release 2/92 (1992).
5. Synopsys, A. E. D. (2011), 'Digital standard cell library', SAED EDK90 CORE Databook.
6. Mohanty, B. K. and Patel, S. K. (2014), 'Area–delay–power efficient carry-select adder', IEEE transactions on circuits and, B.K. and Patel, S.K. systems II: express briefs 61(6), 418–422.

7. Srivastava, P. and Chung, E. (2022), 'An asynchronous bundled-data barrel shifter design that incorporates a deterministic completion detection technique', IEEE Transactions on Circuits and Systems II: Express Briefs 69(3), 1667–1671.
8. Lai, K. K. (2016), Novel Asynchronous Completion Detection For Arithmetic Datapaths, Taylor's University, Malaysia.

Appendix A
Floating-Point Addition

Floating-point addition is the fundamental floating-point operation. According to [1], 60% of the signal processing algorithms require addition operation for real data application. Oberman [2] has analysed the frequency of floating-point operations in various applications and observed that 55% of instructions can be implemented by floating-point adder, represented by Fig. A.1. It makes the floating-point adder (FPA) an essential building block for various applications like data processing units, DSP, embedded arithmetic processors, etc. Several existing algorithms proposed for FPA have been analysed in [3] and [4]. The basic algorithm of floating-point addition is improved to reduce the latency and power consumption of FPA. The LOP algorithm replaces Leading One Detector (LOD) with Leading One Predictor (LOP), originally developed by Hokenek and Montoyc [5], and it is used by Quach and Flynn [6], Suzuki et al. [7], Malik and Ko [8] among others. Another alternative approach known as FAR/CLOSE path algorithm is introduced by Farmwald [9], and its design has been improved over the years by various researchers.

Floating-point addition requires many different sub-operations, which makes the addition operation complicated [10–12]. For adding two floating-point numbers, exponents of both the operands must be equal. If the exponents are unequal, the mantissa of the smaller exponent is shifted to the right, and the smaller exponent is incremented for each shift until it is equal to the larger exponent. After the exponent matching, the mantissa of both the operands is added by using a fixed-point adder, and the exponent value of the result is equal to the larger exponent. The result is then normalised if necessary, to convert it into IEEE 754 format. This is the standard algorithm to perform addition operation on two floating-point numbers. Various addition algorithms for the floating-point number have been compared in [3], and in [13] it was demonstrated that the circuit design can be optimised by selecting appropriate description language. The simulation time and implementation area are affected by the description style and control statement. The single path and double datapath implementation of FPA was discussed in [11]. The concept of Leading Zero Anticipator (LZA) was introduced in [5], which can anticipate the leading zero for

© The Author(s), under exclusive license to Springer Nature Switzerland AG 2022
P. Srivastava, *Completion Detection in Asynchronous Circuits*,
https://doi.org/10.1007/978-3-031-18397-3

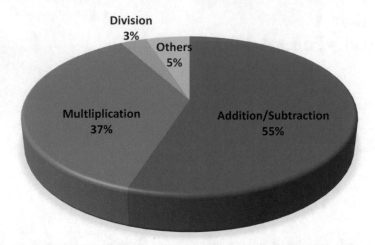

Fig. A.1 Distribution of floating-point instructions

the normalisation of result parallel to the addition operation. Leading One Predicter (LOP) algorithm is proposed by Quach and Flynn [6] for post-normalisation, which is similar to the LZA. The design is further improved by Oberman [14], Oberman et al. [15], Bruguera and Lang [16], and Oberman and Flynn [17]. An area and delay-efficient Leading One Detector (LOD) was introduced by Oklobdzija [18] for normalisation of the result. Oberman et al. [15] have proposed a two-path algorithm as they mentioned that 43% of floating-point operations have an exponent difference of either 0 or 1.

Figure A.2 shows the basic algorithm to add/subtract two floating-point numbers A and B, as explained in the following steps, where operands A and B are represented in the IEEE format as $S_A \ E_A \ M_A$ and $S_B \ E_B \ M_B$, respectively.

Step 1: Calculate the exponent difference: $d = |E_A - E_B|$.

Step 2: Alignment: Monitor the carry C_{out} to identify the smaller operand. The smaller operand needs to be shifted by amount d before performing addition and the larger operand is directly fed to the adder.

Step 3: The shifter output is fed to the XOR gates with a Control signal such that: If Control $= 0$, effective operation is addition. The shifted mantissa from Step 2 would be transferred as it is to the adder with $C_{in} = 0$, as performing XOR operation with zero bit would provide the same variable, i.e., $x \oplus 0 = x$.

If Control $= 1$, the effective operation is subtraction. 2's complement of the shifted mantissa from Step 2 would be transferred to the adder, as performing XOR operation with one bit would provide the complement of the variable, i.e., $x \oplus 1 = x'$ and $C_{in} = 1$ would provide the additional 1 for 2's complement. The sign of the result would be calculated by the sign computation block.

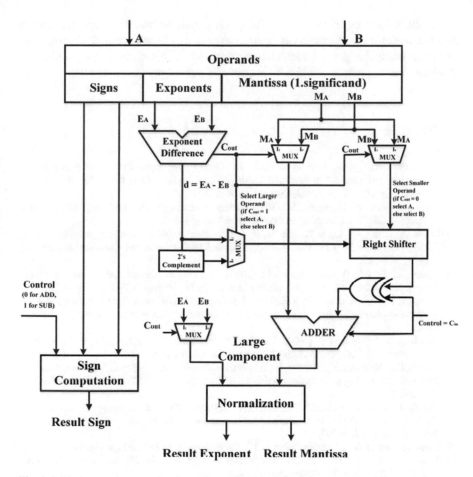

Fig. A.2 Basic asynchronous floating-point adder (AFPA) architecture

Step 4:   Add the shifted mantissa with the mantissa of the other number. The sum-
           mation output is sent to the normalisation unit to provide the final outcome
           in the IEEE format. The normalisation unit performs the following
           operations:

- Convert the resultant mantissa into a signed magnitude format, and find the 2's
  complement of the result if it is negative.
- Leading One Detector (LOD): Detect the leading one in the result after subtrac-
  tion operation, to find the amount required for left shifting.
- For addition operation, either no shifting or maximum 1-bit right shifting is
  required.
- Convert the result into the standard IEEE format.
- Rounding: Round off the bits shifted out due to normalisation.

As discussed above, mantissa shifting for exponent matching and addition of aligned mantissa are two main key operations of FPA. The adder and shifter modules designed using the generic architecture proposed in this study can be utilised to implement an asynchronous FPA to improve its overall performance [19].

References

1. Pappalardo, F., Visalli, G. and Scarana, M. (2004), An application-oriented analysis of power/precision trade-off in fixed and floating-point arithmetic units for VLSI processors, in 'Circuits, Signals, and Systems', Citeseer, pp. 416–421.
2. Oberman, S. F. and Flynn, M. J. (1997), 'Design issues in division and other floating-point operations', IEEE Transactions on Computers 46(2), 154–161.
3. Malik, A., Chen, D., Choi, Y., Lee, M. H. and Ko, S.-B. (2008), 'Design tradeoff analysis of floating-point adders in FPGAs', Canadian Journal of Electrical and Computer Engineering 33(3/4), 169–175.
4. Daoud, L., Zydek, D. and Selvaraj, H. (2015), A survey on design and implementation of floating-point adder in FPGA, in 'Progress in Systems Engineering', Springer, pp. 885–892.
5. Hokenek, E. and Montoye, R. K. (1990), 'Leading-zero anticipator (LZA) in the IBM RISC system/6000 floating-point execution unit', IBM Journal of Research and Development 34(1), 71–77.
6. Quach, N. and Flynn, M. J. (1991), Leading One Prediction-Implementation, Generalization, and Application, Computer Systems Laboratory, Stanford University.
7. Suzuki, H., Morinaka, H., Makino, H., Nakase, Y., Mashiko, K. and Sumi, T. (1996), 'Leading-zero anticipatory logic for high-speed floating-point addition', IEEE Journal of Solid-State Circuits 31(8), 1157–1164.
8. Malik, A. and Ko, S.-B. (2005), Effective implementation of floating-point adder using pipelined lop in FPGAs, in 'Canadian Conference on Electrical and Computer Engineering, 2005.', IEEE, pp. 706–709.
9. Farmwald, M. (1981), 'On the design of high-performance digital arithmetic units'.
10. Waser, S. and Flynn, M. J. (1982), Introduction to Arithmetic for Digital Systems Designers, Saunders College Publishing/Harcourt Brace.
11. Ercegovac, M. D. and Lang, T. (2004), Digital Arithmetic, Elsevier.
12. Parhami, B. (2010), Computer arithmetic: Algorithms and hardware designs, Vol. 20, Oxford University Press.
13. Chen, R.-D., Chou, Y.-C. and Liu, W.-C. (2011), Comparative design of floating-point arithmetic units using the balsa synthesis system, in '2011 International Symposium on Integrated Circuits', IEEE, pp. 172–175.
14. Oberman, S. F. (1996), Design Issues in High Performance Floating Point Arithmetic Units, Citeseer.
15. Oberman, S. F., Al-Twaijry, H. and Flynn, M. J. (1997), The snap project: Design of floating-point arithmetic units, in 'Proceedings 13th IEEE Symposium on Computer Arithmetic', IEEE, pp. 156–165.
16. Bruguera, J. D. and Lang, T. (1999), 'Leading-one prediction with concurrent position correction', IEEE Transactions on Computers 48(10), 1083–1097.
17. Oberman, S. F. and Flynn, M. (2001), Advanced Computer Arithmetic Design, J. Wiley.
18. Oklobdzija, V. G. (1994), 'An algorithmic and novel design of a leading zero detector circuit: Comparison with logic synthesis', IEEE Transactions on Very Large Scale Integration (VLSI) Systems 2(1), 124–128.
19. Srivastava, P., Chung, E. and Ozana, S. (2020), 'Asynchronous floating-point adders and communication protocols: A survey', Electronics 9(10), 1687.

Appendix B
Shifter Test Bench

sync_35_tb.v

D:/Vivad0Nov20/Barrel_Shifter_Codes/bs_sync_0.35/bs_sync_0.35.srcs/sim_1/new/sync_35_tb.v

```
 1 `timescale 1ps / 1ps
 2 module sync_35_tb;
 3     reg     clk;
 4     reg     rst;
 5     reg     [31:0]a;
 6     reg     [4:0]s;
 7     wire    [31:0]out;
 8     reg     [31:0]expected_data;
 9     bs_sync_35 uut(.out(out),.s(s),.a(a));
10
11     initial begin
12         $display ("\n\synchronous barrel shifter test\n\n");
13         clk = 0; rst = 0;
14         #1130 rst = 1; #1500 rst = 0;
15         #100000 $finish; // adjust in order to capture the full simulation
16     end // initial
17
18     // clock signal
19     always #1180 clk = !clk; //its working on 1180, but failing on 1179
20                             //clock period=2360ps at least
21     // test data registers
22     always @(posedge clk)
23     begin
24         if (rst)
25           begin
26             a = 32'hx4b20fd35; s = 0; expected_data = 32'hx4b20fd35;
27           end
28         else
29           begin
30             if (out != expected_data)
31             $display("Failed at %t %h>>%d = %h (expected %h)\n",
32             $time, a, s, out, expected_data);
33             a = a + s;
34             s = s + 1;
35             expected_data = a>>s;
36           end // if
37     end // always
38 endmodule
```

Fig. B.1 CBS test bench

```
async_35_tb.v
     D:/Vivad0Nov20/Barrel_Shifter_Codes/bs_async_35/bs_async_35.srcs/sim_1/new/async_35_tb.v
   1        `timescale 1ps / 1ps
   2      module async_35_tb;
   3           reg [31:0]a;
   4           reg [4:0]s;
   5           reg req;
   6           wire ack;
   7           wire [31:0]out;
   8           reg  [31:0]expected_data;
   9           bs_async_35 uut (.out(out),.ack(ack),.s(s),.a(a),.req(req));
  10
  11           initial
  12           begin
  13            $display ("\n\asynchronous barrel shifter\n\n");
  14            // initialize inputs to UUT
  15            req = 0; a = 32'hx4b20fd35; s = 0;
  16 O         // initialise/reset UUT
  17           // req = 0 will force the delay model to produce an ack = 0
  18           // also let valid data propagate thru' UUT to initialise it
  19 O          @(negedge ack)
  20 O           begin
  21 O            #1 assign req = 1'b1;
  22            $display("Display_1 %t req=%d ack=%d",$time, req,ack);
  23           end
  24          @(posedge ack)
  25          begin
  26            $display("Display_2 %t req=%d ack=%d",$time, req,ack);
  27            #1 assign req = 1'b0;
  28 O          $display("Display_3 %t req=%d ack=%d",$time, req,ack);
  29          end
```

Fig. B.2 ABBS test bench

```
31      for (s = 0; s <= 32; s = s + 1)
32        begin
33            a=a+s;
34          assign expected_data = a>>s;
35          @(negedge ack)
36          begin
37              #1
38              assign req = 1'b1;
39                $display("Display_4 %t req=%d ack=%d",$time, req,ack);
40          end
41          @(posedge ack)
42          begin
43              //$display("Display_5 %t req=%d ack=%d",$time, req,ack);
44              #1
45              assign req = 1'b0;
46              $display("Display_6 %t req=%d ack=%d",$time, req,ack);
47              if (out != expected_data)
48                  $display("Failed at %t %h >> %d = %h (expected %h)",
49                  $time, a, s, out, expected_data);
50              else
51                  $display("Shifted at %t %h >> %d = %h (expected %h)",
52                  $time, a, s, out, expected_data);
53          end
54      end // for
55      $finish;
56      end // initial
57  endmodule
58
```

Fig. B.2 (continued)

Index

Printed in the United States
by Baker & Taylor Publisher Services